NORTH CAROLINA
STATE BOARD OF COMMUNITY COLLEGES
LIBRARIES
ASHEVILLE-BUNCOMBE TECHNICAL COMMUNITY COLLEGE

DISCARDED

JUN 2 3 2025

The Engineer in History

Worcester Polytechnic Institute
Studies in Science, Technology and Culture

Lance Schachterle and Francis C. Lutz
Co-Editors

Vol. 14

PETER LANG
New York • San Francisco • Bern • Baltimore
Frankfurt am Main • Berlin • Wien • Paris

John Rae
Rudi Volti

The Engineer in History

PETER LANG
New York • San Francisco • Bern • Baltimore
Frankfurt am Main • Berlin • Wien • Paris

Library of Congress Cataloging-in-Publication Data

Rae, John.
　　The engineer in history / John Rae and Rudi Volti.
　　　p.　　cm. — (Worcester Polytechnic Institute Studies in
　science, technology, and culture; vol. 14)
　　　Includes bibliographical references and index.
　　　1. Engineering—History.　2. Engineers—History.　I. Volti, Rudi.
　II. Title.　III. Series.
　TA15.R33　　　1993　　　　620.009—dc20　　　　93-22895
　ISBN 0-8204-2062-X　　　　　　　　　　　　　　　　　　CIP
　ISSN 0897-926X

The paper in this book meets the guidelines for permanence and durability of the Committee on Production Guidelines for Book Longevity of the Council on Library Resources.

© Peter Lang Publishing, Inc., New York 1993

All rights reserved.
Reprint or reproduction, even partially, in all forms such as microfilm, xerography, microfiche, microcard, offset strictly prohibited.

Printed in the United States of America.

Table of Contents

1	**Introduction**	1
2	**The Ancient World**	7
	Engineers in Ancient Egypt	8
	The Engineers of Ancient Greece	10
	Engineers and Engineering in Ancient China	17
	The Engineers of Rome	19
	The Ancient Engineers in Perspective	25
3	**The Medieval Engineer**	35
	An Era of Innovation	36
	The Master Mason and His Work	38
	The Design Techniques of the Medieval Engineer	43
	The Ideas and Accomplishments of One Medieval Architect-Engineer	45
	Aeronautical Engineering in the Middle Ages	46
	The Summit of Medieval Engineering: The Clock	47
	The Military Engineer	50
	Medieval Engineering in Transition	52
4	**The Renaissance: The Recognition of the Engineer**	57
	Engineers of the Early Renaissance	59
	Leonardo da Vinci	62
	The Efflorescence of Italian Engineering	65
	The Engineers of Northern Europe	70
	The Mining Engineer in Germany	73
	Engineers in France	74
	The Legacy of the Renaissance Engineers	79

5   The Emergence of the Civil Engineer 85
      The Organization of Engineering in France 86
      The Civil Engineer in Britain 90
      Britain and France Compared 97
      The Railroad Engineers 99

6   The Mechanical Engineer and the Industrial Revolution 109
      The Introduction of Steam Power 112
      The Machine Makers 121
      The Social Origins of the British Engineers 129

7   The Founding of American Engineering 137
      Early American Engineers 139
      Civil Engineers 142
      Mechanical Engineers 150

8   Into the Twentieth Century: Engineers Join the Organizational Revolution 159
      The Growth of Engineering Specializations 163
      The Industrial Research Laboratory 164
      The Engineer and the Organization: Conflict and Concord 167
      The Ascent of Japanese Engineering Exemplified 170
      The Working Life of Engineers 172

9   Engineers as Managers 183
      Engineers, Owners, and Managers 183
      Engineers as Managers: The Soviet Union 189
      Scientific Management 192
      Technocracy 196
      Engineers and Modern Management 199

10   Educating Engineers 207
      The Development of Engineering Education in Europe 207
      Engineering Education in the United States 212

Educating Engineers in Asia 216
Professional Societies and the Upgrading of
 Engineering Education in the United States 218
Graduate Education in Engineering 221
The Consequences of Modern Engineering
 Education 223
Engineering Education and the Quest for
 Professionalization 225

11 The Engineer Today: Origins, Roles, and Status......... 235
The Social Origins of Engineers 236
Women Engineers 238
The Search for Professional Status 240
The Engineer as Whistle Blower 244
Professional Societies and Professional Ethics.............. 248
An Inclusive Occupation 251
The Managerial Track as an Alternative to
 Professionalization 254
Facing the Future 255

Index.. 263

Chapter 1

Introduction

Only the most ardent technophile would judge a civilization solely by the level of its technology. Numerous depressing examples from our own century remind us that sophisticated technologies can be used for appalling purposes. More generally, a high level of technological development can co-exist with debased morals, repressive governance, and philistine artistic styles. But it would be no less an error to assume that technology is irrelevant to social and cultural advance. It is no exaggeration to say that technology provides the foundations of civilization. Not only does technology allow us to meet our material needs, it also allows us to transcend mere existence by directly and indirectly promoting the advance of science, the arts, and all the other elements of civilized life. At the same time, many technological achievements do more than simply make the development of these cultural spheres possible. Technology's products can have an aesthetic value of their own; a well-designed bridge can appeal to the artistic spirit as much as a literary or musical masterpiece.

Technology does not create itself; it requires the efforts of the technologist. Many occupational roles can be subsumed under this title—craftsman, organizer, researcher, and skilled worker. In these pages we will consider the central player in this cast, the engineer. Linguistically, the term "engineer" is a comparatively recent arrival in history. In antiquity, the designer and builder of a temple, palace, fortification, harbor, road, or water supply system was known as an *architekton* in Greece and an *architectus* in Rome. Like modern architects, they were concerned with structures, but on a broader scope. Tertullian, one of the early Church Fathers, seems to have applied the term "ingenium" (ingenious device) to siege artillery about 200 A.D.[1] However, the derivative word "ingeniator," the person who makes the ingenious device, does not appear until much later; only in the late Middle Ages and early Renaissance do we begin to find vernacular forms such as "ingenieur" and "engineer". For a long time thereafter the word continued to designate the designers and builders of military structures and devices, as can be seen in the memorable lines of Shakespeare, "For 'tis the sport to have the engineer hoist with his own petar."[2] It was not until the eighteenth century that the term "civil engineering" emerged, referring specifically to non-martial pursuits.

The word "engineer" has been an elastic one, encompassing everyone from the operators of locomotives to policy-makers bent on the promotion of "social engineering." A more restrictive definition, and one that encompasses the subjects of this book, is offered by Samuel Florman, who describes engineering as the "art or science of making practical application of the knowledge of the pure sciences."[3] This definition is acceptable as long as it is understood that "science" should be used broadly, in the sense of organized, empirically derived knowledge. Through most of history scientific knowledge was applied intuitively and was gained through experience rather than through academic channels. It also must be emphasized that a considerable amount of good, practical engineering has been done in complete ignorance of underlying scientific principles, or through the application of principles that were simply wrong.

It cannot be denied that engineering and science have moved ever closer together; certainly there seems to be nothing in the distant past comparable to the intimate (but highly complex and often misunderstood) relationship that has developed between science and engineering over the past century or two. Even so, it is hard to make a completely confident assessment of the science-engineering relationship of earlier times. We do not know enough about about the education or training of early engineers to make any accurate assessment of how much they know of the science and mathematics of their times. The engineer did not (and still does not) need to have a formal education in science, for the necessary information could have been acquired through actual practice or transmitted by word of mouth in a master-apprentice relationship, with no surviving record for the benefit of the historian.

The fragmentary records of early civilizations may well have created a misleading picture of the engineers who provided the technology on which those civilizations rested. The surviving records were the work of the literati, who in most periods of history have considered technology to be unworthy of mention if not positively distasteful. The gulf between the humanities on the one hand and science and technology on the other is hardly a recent phenomenon. In most early civilizations (and a good many later ones) the cultural and economic elite regarded manual labor as suitable only for menials and slaves, and the practical application of knowledge as a debasement of that knowledge. As a result, the individuals who did the engineering work were likely to be regarded as inferiors and therefore ignored by chroniclers drawn from the ranks of the elite.

Over the centuries, few practicing engineers were products of the Academy; indeed, even today the engineer is likely to be looked upon as an interloper in the sacred groves. But all engineers have received some sort of education, and engineering practices have always been tightly interwoven with modes of education. We will chart this relationship throughout this book.

In writing a history of engineers we have to resign ourselves to having far less biographical material than we would like. The records of history tell us much more about about engineering achievements than they do about the people who created them. Even the records of technology itself, the structures and the artifacts, are far more likely to commemorate emperors, kings, governors, mayors, and lesser political worthies than the engineers who really were responsible. A classic example appears in American history. DeWitt Clinton is acclaimed in all the standard texts as the "builder" of the Erie Canal. Admittedly, the canal might not have been built without Clinton's foresight and promotional efforts, but he was not its builder. The surveying, land clearing, excavating, bed construction, and lock building were the work of James Geddes, Benjamin Wright, and Canvass White, a threesome known only to a limited circle of scholars.

The purpose of this book is to add to our understanding of these people who have done so much while working in relative anonymity. We will be concerned with their social origins, the ways in which they were educated, their relationships with employers and patrons, and how they were regarded in their communities and societies. Our overall framework is a historical one, and we have generally followed a chronological approach, using conventional periodizations such as Antiquity, the Renaissance, the Industrial Revolution, and so forth. When we get to more recent times, roughly the period following the consolidation of the Industrial Revolution, we depart somewhat from a chronological presentation and take on a more thematic approach. In part, this reflects the greater abundance of historical and sociological material available to us, which allows a more extensive and, it is to be hoped, sophisticated discussion of many of the topics explored in earlier chapters. In these latter chapters we also spend a bit less time on individual engineers and more on the general social contours of engineering. To some extent this is necessitated by the tremendous expansion of the number of engineers that has taken place in the last hundred and fifty years, for it is far more difficult to select individual engineers whose work has been preeminent in the shaping of technology. It also reflects the fact that engineering has been strongly influenced by fundamental economic and social trends: growing specialization, the expansion of educational institutions, the power of large and complex organizations, and the efforts of many occupational groups to secure professional status for their members.

No pretense is made that this book offers an encyclopedic coverage of engineers and their works.[4] Many renowned engineers have been left out, and the ones who are featured often receive schematic treatment. Moreover, many engineering specialties get limited coverage at best. This may be considered a shortcoming, but we hope not a serious one. Rather than attempting the nearly impossible task of providing complete coverage, we have chosen to explore engineering over the ages through a narration of how the lives and accomplishments of particular engineers have been shaped by the societies in which they lived and worked. Most of the engineers highlighted in this book are important in their own right, but equally important, they illustrate important aspects of engineering, some of them universal, others limited to particular times and places. We hope that the engineer will emerge from these pages as both a historic and a historical figure.

Acknowledgements

This book covers a fair amount of ground, and on several occasions our expertise has been stretched by the requirements of the narrative. We are appreciative of the help provided by Bradford Blaine, Francis Lutz, Florence Rae, Donald Remer, Robert Schofield, Ann Stromberg, and John Truxal, who read portions of the text and offered helpful suggestions. They of course are not responsible for any errors of fact or interpretation to be found in this book.

***

Since this is a co-authored work, the use of the word "we" in various parts of book is not simply a stylistic convention. At this point, however, one of the authors must speak only for himself, for the final draft of this book was prepared after the death of John Rae on 24 October 1988. I can only hope that this book would have met with his approval, and that I have not deviated too far from his ideas and intentions.

Notes

1 James K. Finch, *Engineering and Western Civilization* (New York: McGraw-Hill, 1951), p. 22.

1 *Hamlet*, Act 3, Scene 4. A petar (or petard) was a small bomb used in siege warfare.

3 Samuel C. Florman, *The Existential Pleasures of Engineering* (New York: St. Martin's, 1976), p. x.

4 For biographical sketches of particularly accomplished engineers, see Roland Turner and Steven L. Gulden, *Great Engineers and Pioneers of Technology, vol. 1: From Antiquity through the Industrial Revolution* (New York: St. Martin's Press, 1982) and James Cavill, *Great Names in Engineering* (London and Boston: Butterworths, 1981). Some useful analytical biographies appear in Terry S. Reynolds, *The Engineer in America* (Chicago: University of Chicago Press, 1991). For a variety of perspectives on the actual work of engineering, see Richard L. Meehan, *Getting Sued and Other Tales of Engineering* (Cambridge, Massachusetts: MIT Press, 1981), Henry Petrovski, *To Engineer Is Human: The Role of Failure in Successful Design* (New York: St. Martin's Press, 1985), and Walter G. Vincenti, *What Engineers Know and How They Know It: Analytical Studies from Aeronautical History* (Baltimore and London: Johns Hopkins University Press, 1990).

Chapter 2

The Ancient World

Most studies of Antiquity have concentrated on the literary, artistic, and philosophical accomplishments of the ancient world. Far less attention is customarily devoted to the development of technology. This is understandable, given the slow pace at which technology advanced in the ancient world. Still, the development and application of new ways of doing things, coupled with the systematic application of existing ones, were essential to the civilizations of antiquity. Most of the technologies that were created and used were the legacy of anonymous workers and artisans, but significant contributions were also made by people whom we would readily identify as engineers. At the same time, the vast majority of them are every bit as anonymous as less exalted workers. The works they left behind provide impressive evidence of their achievements, but we know very little about who they were, how they were trained, and how they went about their work. What one frustrated scholar of Rome said can be applied to all: "At present we know virtually nothing about Roman engineers."[1] Names occasionally appear, sometimes with fragments of information, but there is never enough to tell us as much as we would like to know about these ancient engineers.

Although much remains unknown, it is evident that some features of engineering have remained unchanged over the centuries. A reading of Emperor Sennacherib of Assyria's egocentric description of the works completed during his reign would likely put a wry smile on the face of a modern civil engineer:[2]

> To increase the productivity of the low-lying fields, through the high and low grounds I dug with iron pickaxes, I ran a canal; those waters I brought across the plain around Nineveh and made them flow through

the orchards in irrigation ditches... I saw pools and enlarged their narrow sources and turned them into a reservoir. To give these waters a course through the deep mountains, I cut through their difficult places with pickaxes and directed their outflow on to the plain of Nineveh. I strengthened their channels, heaping up their banks mountain-high... with stones I lined the canal and Sennacherib's Channel I called its name.

Sennacherib must have been a very busy man, what with all that wielding of a pickaxe. One wonders how he found the time to attack Judah "Like the wolf on the fold"[3] until his army was destroyed by pestilence. The emperor returned home, perhaps concluding that the pickaxe was, if not mightier, at least safer than the sword. But if he gave to himself all the glory, Sennacherib at least had the grace not to blame the engineers when things went wrong. When the water broke through a gate just before the official opening of the city of Nineveh's water-supply system, he chose to interpret this as a sign of divine favor, and instead of executing his engineers he gave them valuable presents.[4]

Engineers in Ancient Egypt

The earliest record of an individual who took on the role of engineer documents the work of Imhotep, the chief minister for the Egyptian pharoah Zoser (c. 2750 B.C.).[5] Imhotep was hardly a humble workman; he also served as Heliopolitan High Priest, and his father had been the royal architect. He is credited with building the first pyramids, a feat made possible because the Egyptians had by this time developed techniques for cutting and handling large blocks of stone. Equally important was the ability of the pharoah's officials to raise and manage thousands of laborers to transport and lift the blocks during the slack agricultural season from the end of July to the end of October when the Nile was in flood.

In addition to his other duties, Imhotep also served as court physician. His greatest recognition derived from this, for some centuries later he was worshiped as the god of medicine. The fact that Imhotep served as a physician and adminstrator as well as an engineer may cast some doubt on his actual involvement with engineering. Still, serving in a multiplicity of capacities was commonplace in those days, for able individuals who could read, write, and calculate were a minuscule few. It was necessary for them to perform a variety of tasks requiring these skills, and in any event, the body of knowledge they had to work with was still sufficiently limited for them to be reasonably competent in several fields.

While Imhotep came from an aristocratic family, noble birth was not always essential for attaining the rank of Royal Architect. The inscription on the tomb of Nekhebu (c. 2600 B.C.) notes that[6]

> His Majesty found me a common builder; and His Majesty conferred upon me the offices of Inspector of Builders, then Overseer of Builders, and Superintendent of a Guild. And His Majesty conferred upon me the office of King's Architect and Builder, then Royal Architect and Builder under the King's supervision. And His Majesty conferred upon me the offices of Sole Companion, King's Architect and Builder in the Two Houses.

It would be helpful to know what combination of technical capability and personal appeal accounted for Nekhebu's occupational progress. We only know that the rewards doled out by the pharoah were a mixture of the exalted and the commonplace: gold, bread, and beer.[7]

The next Egyptian architect-engineers about whom anything appreciable is known were Ineni and Senumut, whose careers fell approximately in the century between 1550 and 1450 B.C., overlapping in the reign of Queen Hatshepsut (c. 1500-1450 B.C.). As with Imhotep, both were of noble birth, and each held the office of chief architect, among other posts, under several pharoahs. Their works included a variety of construction projects — palaces, tombs, and obelisks. Each left an epitaph extolling his own achievements, free of the slightest trace of inhibition or modesty. Each used the title "foreman of foremen," to which Senumut added "chief of chief of works." Whether they performed as engineers in the modern sense of the word cannot be ascertained, but they certainly took on at least some of the functions of the civil engineer. As Ineni boastfully summed up his career:[8]

> I became great beyond words. I will tell you about it, ye people. Listen and do the good that I did, just like me. I continued powerful in peace and met with no misfortune; my years were spent in gladness. I was neither traitor nor sneak, and I did no wrong whatever. I was foreman of the foremen and did not fail. I never hesitated but always obeyed superior orders, and I never blasphemed sacred things.

The British Egyptologist who translated this epitaph noted that if Ineni did indeed work with Egyptian labor for some forty years and never blasphemed, that was certainly not the least of his accomplishments.

In addition to their more famous building of the pyramids, the Egyptians also built canals around the First and Second Cataracts of the Nile, and as early as 1900 B.C. constructed a canal from the Nile Delta to the Red Sea, probably by way of the Bitter Lakes on the present Suez Canal.[9] Unfortunately, there is no record of who the engineers were or how they went about their work. The Nile-Red Sea Canal fell into disuse; periodic attempts to restore it were made by later pharoahs and even by the Persians and Romans, but they were either short-lived or failed altogether. As with some modern large engineering projects, the whole concept may have been fatally flawed; the weather and the navigational hazards of the Gulf of Suez were too much for Egypt's vessels, so it was not worthwhile to make a continuous effort to keep the canal open.[10]

The Engineers of Ancient Greece

With the Greeks we are somewhat better off— not much, but enough to call into question the conventional assumption that the Greek genius was totally absorbed in abstract philosophy and mathematics, and was unconcerned with practical affairs. It is not difficult to find passages in Socrates and Plato in which practical matters and the work of artisans are subject to scorn. In Aristotle's estimation, for example, "No man can practice virtue who is living the life of a mechanic."[11] Still, the accomplishments of the Greeks in what we would today call civil engineering are hardly negligible. Greek engineers surmounted the intellectual snobbery of the philosophers and created monuments to human intellect no less impressive than the philosophical systems of their contemporaries.

The Greek title for the engineer was *architekton*, master of the practical arts. He combined the functions of the modern architect and civil engineer, and often served as a mechanical engineer for good measure.[12] Like his predecessors in Egypt and Mesopotamia he was concerned with structures of all kinds — palaces, temples, harbor works, fortifications, bridges, and aqueducts, as well as mechanical contrivances such as cranes and siege engines.

Many engineering works combined structural soundness with a keen appreciation of aesthetics. The design of temples and other monumental structures was rationalized by using multiples of standard modules, such as a frieze of triglyphs, for basic dimensions.[13] At the same time, optical illusions were successfully employed to improve the appearance of buildings. Most notably, the architects who designed the Parthenon, Iktinos and Kallikrates, gave its columns a slight outward bulge so that they would appear straight, and cambered its steps so that they would appear level. In translating their designs into actual structures, the architect-engineers employed sets of graduated blocks, templates made from bent rods, and simple leveling devices. It is also likely that they worked closely with a building's contractor in order to work out the constructional and aesthetic problems that emerged in the course of construction.[14]

The design of the Parthenon would today be labelled as architecture rather than engineering, although the distinction between the two was unknown in ancient times. In any event, some works described by the Greek historian Herodotus were engineering projects pure and simple. One was a water-supply system on the island of Samos that required tunneling through a 900-foot hill. Built in the 6th century B.C. by Eupalinos, it was 3,300 feet long, about 5.5 feet wide and deep, and had a trench three to 25 feet deep for the conduit.[15] It was cut from both ends and had a jog in the middle, for the two bores missed by 20 feet horizontally and three feet vertically. This error was unimportant for a tunnel meant only for the transport of water. In view of the elementary instruments and crude tools available to him, the fact that Eupalinos came as close as he did shows that he was a highly competent engineer.

Another achievement noted by Herodotus was the set of bridges that carried the army of the Persian king Xerxes across the Hellespont to invade Greece in 480 B.C. Xerxes employed Phoenician and Egyptian engineers, who built what were essentially pontoon bridges, the Egyptians tying the structure together with papyrus cables while the Phoenicians used flax, but their efforts were ruined by a great storm. Xerxes was not one to accept a structural failure graciously; as Herodotus narrated the subsequent events, Xerxes ordered that the Hellespont receive 300 lashes and have a pair of fetters cast into it. The engineers were even more unfortunate; Xerxes commanded that they have their heads cut off.[16]

The successors to these unfortunates were Greek; at least the principal engineer was a Greek named Harapolos. He also used two floating bridges, but with the example of his predecessors in mind he was undoubtedly more careful with his stress calculations. 360 vessels were linked together to form the upstream bridge and 314 were used for the one downstream, all securely anchored with their prows facing the current and connected by cables.[17] Harapolos, however, avoided the nationalistic prejudices of his predecessors and used two flax and four papyrus cables on each bridge. The crossing was impressive by any standard, and many centuries would pass before anything similar was attempted in the Western world. Everything held together, and Xerxes' host crossed the strait in safety, undoubtedly to wish later that the bridges had never been built, given the results of the ill-fated campaign.

Impressive as they were, the engineering accomplishments of Greece in its Golden Age were in some ways surpassed during the Hellenistic era, when Greek civilization borne by the conquests of Alexander the Great merged with the civilizations of the conquered lands. Ktesibius (or Ctesibius) of Alexandria (fl. 270 B.C.), the son of a barber, devised musical instruments, water clocks, and numerous mechanisms that were pneumatically or hydraulically powered. He is also credited with the invention of the force pump and being the first to employ metal springs. Philon of Byzantium (fl. 250 B.C.), who may have been a student of Ktesibius, also designed toys and gadgets run by air and water. Of greater practical importance, he described and may have contributed to the design of water wheels used for the lifting of water.[18] Philon also wrote several works on what today we would call mechanical engineering, only fragments of which have survived. Significantly, these works were written in vernacular Greek, rather than the Attic language of scholars, a possible indication that he began as an artisan and was not a member of the elite stratum. At the same time, his extended use of mathematics in his treatises set him apart from contemporary workers, and indeed from most engineers until the 19th century.

The most impressive blend of mathematics, science, and engineering talent is to be found in Philon's contemporary, Archimedes (287-212 B.C.). Unfortunately, both ancient and modern assessments tend to stress the first two at the expense of the third. Beyond question, Archimedes was a brilliant scientist and mathematician, the Newton of antiquity. But historians have gone astray in attributing to him a disdain for the practical accomplishments of engineering, often quoting the Roman historian Plutarch's assertion that Archimedes "...would not consent to leave behind him any treatise on [his inventions], but regarding the work of the engineer and every art that ministers to the needs of life as ignoble and vulgar, he devoted his earnest efforts only to those studies the subtlety and charm of which are not affected by the claims of necessity."[19]

Recent scholarship has strongly challenged this view, arguing that Archimedes was an engineer first, and a scientist and mathematician second.[20] He was a protege, and likely a relative, of King Hieron of Syracuse (c. 270-215 B.C.). Hieron sent the young Archimedes to Alexandria to complete his education, and upon his return put him in charge of the royal shipyards, an important export industry of Syracuse. Archimedes did calculations of centers of gravity and buoyancy, mathematical activities definitely intended for practical application. He also designed and supervised the building of ships, his principal achievement being a 400 foot-long combination warship, merchant vessel, and pleasure craft, a ship so large that it had to be launched by compound pulleys. Only Syracuse and Alexandria had harbors capable of accommodating this ship, and the seagoing white elephant was eventually given to the ruler of Egypt, Ptolemy Philadephos (285-246 B.C.)

Since ship design and ship building are undoubtedly engineering, there can be no denying that in addition to his other accomplishments Archimedes was by definition an engineer. It is even possible that his shipyard studies of buoyancy were the real source of his famous solution to measuring the gold content of the king's crown, rather than his legendary plunge into a bathtub. If so, it was one of the first recorded instances of scientific knowledge being derived from engineering rather than vice-versa.

Archimedes is supposed to have been especially ashamed of the military engines he devised in the defense of his city. As with his alleged views on the vulgarity of engineering, this notion is based largely, if not exclusively, on Plutarch's account of the siege of Syracuse in his biography of Marcellus. Plutarch was a colorful writer, but not necessarily an accurate historian. His description of the engines is clearly hearsay, and he provides no comprehensible description of them. The evidence offered for Archimedes' alleged contempt for his own handiwork is that he never wrote anything about it. This is hardly conclusive, for there is no telling how many of Archimedes' writings have been lost. Plutarch also notes that "...all the rest of the Syracusans were but a body for the designs of Archimedes and his the one soul moving and managing everything; for all other weapons lay idle, and his alone were employed by the city both in offense and defense."[21] This is surely an exaggeration, but it hardly indicates a diffident approach to military engineering on Archimedes' part. As with the earlier Greek philosophers, it is likely that Plutarch attributed to Archimedes the attitude toward detached, abstract speculation that Plutarch himself thought proper, and later writers have accepted this version uncritically.

At the time of Archimedes a far less distinguished Egyptian Greek named Cleon served as an engineer during the reign of Ptolemy Philadephos. We happen to know something about him because bits and pieces of his official documents were found in the late 19th century by Sir Flinders Petrie, and compiled a few years later by a French scholar.[22] There was nothing about Cleon's career that was very distinguished, but it does provide us some insight into the nature of engineering long ago. He today would be termed a district engineer, as he was in charge of a section of the elaborate irrigation system developed under the Ptolemies in the area of present-day Fayum. He was obviously a member of an elaborate bureaucratic structure and presumably had risen through the ranks, but we know nothing of his training as an engineer or how he entered the system.

Cleon supervised the entire water supply of a substantial district that took water from the Nile and channeled it into a previously semi-desert area. He was responsible for maintaining canals and dikes, operating sluices, and overseeing all the operations the system required. Much of his work was necessarily managerial; he had under him a deputy and several junior engineers, along with quarrymen, masons, teamsters, and other craftsmen and laborers. He also dealt with independent contractors. Much of the surviving documentation is concerned with matters that would evoke sympathy from any present-day engineer: squabbles over prices and wages, materials that were slow to arrive, complaints that the flow of water was too much or too little, and, inevitably, appeals for prompter payment of outstanding bills.

As sometimes happens with well-entrenched administrators, Cleon eventually got careless. He was charged with misuse of funds and removed from his post. There is not enough information to judge the merits of the case; his biographer says "Cleon had so many activities on his hands that he could have been negligent without being guilty."[23]

His misfortune at least lets us see something of the family life of one engineer 22 centuries ago. Cleon had two sons, the younger a student at home, and the older apparently seeking an official position; the first of his extant letters has the son writing home for money. Cleon's wife, Metrodora, wrote while his troubles were unfolding, "You would have wished me to be with you, and I would have left everything to go there."[24] The older son offered his advice and assistance, by this time apparently having advanced beyond his earlier financial embarrassment. Cleon still lost his job, but he seems to have enjoyed a peaceful retirement.

Egypt under the Ptolemies was a great center of Hellenistic culture, with the Museum and Library of Alexandria at its focus. Although primarily a center for detached scholarship, it did provide some support for technological endeavors. Its best-known member was Heron (or Hero), who lived from about 50 A.D. to about 120 A.D. He wrote extensively on technical matters, and happily many of his texts have survived. His writings describe mechanisms and devices created by Ktesibius and Philon as well as some that were his own creations. Among other things, he devised a hydraulic fire extinguisher, a coin-operated holy water dispenser, a device for opening temple doors by air pressure, and an improved surveyor's transit. Heron is also famous in the history of technology for his invention of a device known as the aeolipile. This was a spherical vessel with two protruding radial tubes that ended in right angles set in opposite directions. When steam entered the vessel and was forced through the tubes, the whole apparatus rotated rapidly. Some historians have hailed Heron as the inventor of the first steam engine, but this seems to overstate the case. The aeolipile was anything but a fully developed mechanism; for one thing it must have been very difficult if not impossible to get a good seal at the pivot where the steam enters the vessel. It is also hard to see how it could be used to perform any useful work; its high rotational speeds would have necessitated some form of step-down gearing, and this would have produced a lot of friction, a fatal flaw for an engine that probably worked at only 1 percent efficiency.[25] The aeolipile was never anything but a toy, and like many of the devices produced by Heron and his precursors it was never employed for productive purposes.

Engineers and Engineering in Ancient China

In the making of siege weapons and other implements of warfare, no ancient civilization equalled that of China. Everyone knows of China's invention of gunpowder, but there is less awareness of China's matchless innovative skill in the creation of crossbows, catapults, and aerial bombs.[26] Civilian technology was no less impressive. Along with the familiar examples of paper, the compass, and porcelain, Chinese accomplishments include ships with watertight bulkheads, the segmental arch bridge, double-acting piston bellows, chain-drive transmission, the wheelbarrow, the spinning wheel, and many others. All of these preceded similar inventions in the West, often by many centuries. Indeed, many epochal Western inventions may in fact have been borrowed from China.

While many Chinese inventions were produced by unknown artisans to meet their own needs, the government could also be an important patron of invention, and many Chinese engineers were in the employ of the state. Throughout history one of the most urgent requirements of the government was the regulation of water supplies, for the Chinese economy and society ultimately rested upon irrigated agriculture. This in return required a fair degree of political centralization to undertake water projects and to administer the distribution of water over large areas. So important was this task that some historians have attributed China's establishment of political centralized government to the imperatives of hydraulic engineering. At the same time, the government had a keen interest in the construction and maintenance of canals and other inland waterways, for these were essential for the transport of taxes (in the form of grain) from the hinterlands to the administrative centers. In fact, it has been argued that the government was more concerned with this aspect of hydraulic engineering than they were with meeting agricultural needs.[27]

Under these circumstances, it is not surprising that some Chinese officials took on a large engineering role. One early engineer-official was Li Bing. In 256 B.C. Li was appointed chief of Shu prefecture in the center of today's Sichuan Province. His major task was to bring the Min River under control in order to break the destructive cycle of floods and droughts that had plagued the region. Li realized that some sort of dam was needed, as well as a way of transporting excess water to an area behind a mountain that blocked the river. To accomplish the first task, Li organized more than 10,000 workers to bore a tunnel 20 meters in diameter through the mountain. The river itself was divided in two by damming it in the middle, with one half going through the newly constructed tunnel. After a number of failed attempts the damming was accomplished by filling bamboo containers with rocks and placing them in the middle of the river. Li subsequently added another dam and spillway, and oversaw the temporary damming of each of the river's branches for annual dredging. The project still functions today, serving 200,000 hectares of cultivated fields.[28]

Military imperatives were also an important source of hydraulic engineering projects. During the third century B.C. Qin Shi Huangdi, the first unifier of China, charged an engineer named Shi Lu with the construction of a canal to move supplies to troops in the South. The task here was to connect two rivers that flowed in opposite directions. In order to do this, Shi canalized the Li River to regulate it. He also built a mound in the middle of the Xiang river to divide its water, and constructed a series of spillways that safely diverted a portion of the river into a canal that led into the Li River. The linking of these rivers allowed navigation for a straight-line distance of 1250 miles, making it possible to travel by water from the latitude of Beijing all the way to the city of Guangzhou, far to the south.[29]

Other Chinese bureaucrats were noted for their engineering accomplishments. A government official by the name of Chao Kuo is credited with the invention of the seed-drill in 85 B.C., a device that was unknown in Europe until the sixteenth century A.D. The production of cast iron was given a boost by the invention of the water-powered bellows by a government official named Du Shi in 31 A.D., and further popularized by another official named Han Qi during the third century A.D.

It is of course possible, even likely, that the actual engineering was the work of artisans in the officials' service, but if nothing else, their patronage did facilitate the development and spread of new technologies. There certainly was an abundance of engineering talent waiting to be tapped, as exemplified by a commoner named Bi Lin, who during the 3rd century A.D. built water supply system for the city of Loyang that used square-pallet chain-pumps and norias.[30] Examples can even be found of slaves serving as engineers. The most notable was Geng Xun (fl. 593 A.D.), who after leading a failed local uprising was made an imperial slave. Benefiting from a personal connection with the Astronomer-Royal, he set to work building a water-driven armillary sphere. His success eventually led to his freedom and a position as Acting Executive Assistant in the Bureau of Astronomy.[31]

Chinese engineers produced many impressive technological accomplishments during ancient times, but as we shall see, the full scope of their talents was to become even more evident in later centuries, when Chinese technical progress stood in sharp contrast to the slow pace of invention in Europe during much of the Middle Ages.

The Engineers of Rome

We will return to China in the next chapter. But first we need to consider the lives and work of those Roman engineers who did so much and about whom we know so little. The evidence of their work remains plainly visible in Europe, North Africa, and the Middle East: aqueducts, structures, remains of roads. Their creators remain tantalizingly obscure. Appius Claudius appears in the historical record as the earliest of the great Roman engineers because as Consul about 300 B.C. he was responsible for building the first of the historic Roman roads, the Appian Way, as well as the first aqueduct supplying Rome with water.[32] He may have been the builder only in the sense that Governor DeWitt Clinton is credited with the building of the Erie Canal, but he could well have been more directly responsible. Roman officials were expected to take direct charge of the construction of public works, and the great majority would have enough military service to give them some practical engineering training.

Every student of Latin knows how Julius Caesar's army built a bridge across the Rhine, although one attempting to build a bridge based on his narrative would end up with a structure more like a dam than a bridge. It was common practice to use soldiers on construction work in peacetime; Roman legions were responsible for 128 construction projects in Britain alone.[33] This practice is strikingly illustrated in one of the few records we have of a professional Roman engineer— in this case the inscription on his tombstone. His name was Nonius Datus, "a veteran of the Third Augustan Legion," who died in what is now Algeria in 152 A.D.[34] Employed to survey an aqueduct and tunnel to supply water for Saldae (now the city of Bougie in Tunisia), he recorded that he marked off the alignment, and "When I had assigned the workers, so that they might each of them know the digging he was to do, I instituted a contest between the marines and the troops from Gaul." The contestants may have gotten carried away, because four years later he was called back to rectify an error in digging that had the two headings hopelessly out of alignment with each other.

Perhaps the greatest of the Roman administrator-engineers was Marcus Vipsanius Agrippa (63-12 B.C.), the trusted lieutenant of Emperor Augustus. Born to a family of no known distinction, he was a close personal friend of Augustus. He first served as a distinguished military commander in Gaul and Germany. His engineering skills may have contributed to his success, for he invented a collapsible tower from which to fire missiles, as well as a catapult-launched grapnel to aid in the forcible boarding of enemy ships. Back in Rome he oversaw the repair and expansion of the municipal water works, and after another successful stint as military commander (this time against the forces of Marcus Antonius), he built the first public bath in Rome, a bridge across the Tiber, and a series of temples. The portico of one of these was eventually incorporated into the Pantheon, one of greatest glories of ancient Rome. He also left behind a much less laudable legacy, for he was the grandfather of the mad emperor Caligula, and the great-grandfather of Nero.[35]

An engineer of a different sort was Sextus Julius Frontinus (35-104 A.D.), who served as water commissioner of Rome under the Emperors Nerva and Trajan. He was born to a noble family (although likely an undistinguished and minor branch), rose through the government bureaucracy, and served as governor of Britain for a few years. In 97 A.D. he took charge of the Roman water supply; it was while serving in this capacity that he made his greatest contributions. He took an active interest in the technical aspects of his job, feeling that expertise should not be delegated to underlings:[36]

> I have always made it my principle, considering it to be something of prime importance, to have a complete understanding of what I have taken on. For I do not think that there is any surer foundation for any kind of undertaking, or any other way of knowing what to do or what to avoid; nor is there anything more degrading for a man of self-respect than to have to rely on the advice of his subordinates in carrying out the commission entrusted to him.

Frontinus' eagerness to involve himself with technical details does not mean that he had little help; in fact he supervised a staff of engineers, surveyors, and clerks, as well as 700 slaves who worked as inspectors, foremen, masons, plumbers, and plasterers.[37] This assemblage was charged with maintaining the elaborate system that kept Rome supplied with water. Of no less importance, Frontinus made a vigorous if not wholly successful effort to prevent well-to-do citizens from tapping into the system and diverting water for their private use (some private use was permitted, for a fee, but the privilege was frequently abused).

To carry out his plans, Frontinus attempted with little success to measure the flow of water. Here he was well behind the designer of the famous aqueduct of Nimes in Gaul. The designer of this masterpiece of Roman civil engineering is unknown, but there is some evidence that it may have been the aforementioned Marcus Agrippa. Whoever he was, the engineer who designed the aqueduct seems to have made use of a principle rediscovered in the 19th century, that the rate of flow can be ascertained if the head (the difference in water level between the channel and the reservoir) and the size of the orifice between the two are known. To determine these, he employed a pair of sluice gates to set the head of the incoming water and the underwater orifice through which it flowed. Measurements of the two could then be used to determine the flow of water into its terminal reservoir.[38]

Although Frontinus' technical acumen did not equal that of the the engineer who designed the aqueduct of Nimes, he was justifiably proud of his work. With a classic flourish of Roman pragmatism he boasted that: "With such an array of indispensable structures carrying so many waters, compare if you will the idle Pyramids or the useless, though famous, works of the Greeks."[39] Given the impressive accomplishments of Roman hydraulic engineering, he can surely be forgiven this rhetorical excess.

While we know most Roman engineers only through their surviving aqueducts, roads, and structures, one Roman engineer has gained a measure of immortality through a book he wrote. In the first century B.C. Marcus Vitruvius Pollio wrote *De Architectura*, basing much of it on Greek texts which have long since disappeared. In addition to providing a great deal of information about design and constructional methods, his book offers many insights into the working lives of Roman engineer-architects. Vitruvius was highly critical of incompetent architects who "beg and wrangle to obtain commissions."[40] They were likely to be "ignorant not only of architecture, but even of construction... falsely called architects in the absence of genuine training."[41]

For Vitruvius, a good architect had to combine theory and practice:[42]

> ...architects who without culture aim at manual skill cannot gain a prestige corresponding to their labours, while those who trust to theory and literature obviously follow a shadow and not reality. But those who have mastered both, like men equipped in full armor, soon acquire influence and attain their purpose.

At the same time, the knowledge and skills required of an architect were many and varied:[43]

> He must have both a natural gift and also readiness to learn. (For neither talent without instruction nor instruction without talent can produce the perfect craftsman.) He should be a man of letters, a skillful draftsman, a mathematician, familiar with historical studies, a diligent student of philosophy, acquainted with music, not ignorant of medicine, learned in the responses of jurisconsults, familiar with astronomy and astronomical calculations.

This is quite a list. One wonders how many practicing engineer-architects of the era measured up; it would certainly be a challenge for today's practitioners. But Vitruvius did not expect mastery of all these subjects, only that the architect-engineer have sufficient understanding to do the best work possible. Thus a knowledge of medicine is required so that buildings can be sited in accordance with physiological needs for light, air, and water, while an understanding of law is necessary for dealing with such matters as water supplies, walls, and contracts. Mathematics is necessary for such varied tasks as laying out building sites, estimating costs, measuring, and solving "the difficult problems of symmetry."

Vitruvius also makes it clear that a knowledge of financial matters is essential to the architect-engineer's successful practice. He is careful to note that "I have not studied with the view of making money by my profession; rather have I held that a slight fortune with good repute is to be pursued more than abounding wealth accompanied by disgrace."[44] But at the same time he makes it clear that along with being a competent designer, a good architect-engineer has to be able to accurately estimate costs and stay within a budget. Vitruvius notes with approval the law of the Greek colonial city of Ephesus, and expresses the wish that it were applied in Rome for both public and private projects:[45]

> ....when an architect undertakes the erection of a public work, he estimates at what cost it will be. The estimate is furnished, and his property is assigned to the magistrate until the work is finished. On completion, when the cost answers to the contract, he is rewarded by a decree

in his honour. If not more than a fourth part has to be added to the estimate, the state pays it and the architect is not mulcted. But if more than a fourth extra is spent carrying out the work, the additional sum is exacted from the architect's property.

Such a law would presumably throttle the charlatans that raised his ire and undermined the status of true architect-engineers. One of Vitruvius' major concerns is that the *architectus* enjoy a place in society commensurate with his talents and accomplishments. Such was not always the case in ancient Rome. There were times when the life of the architect-engineer could be downright perilous, as when, according to one story, the Emperor Hadrian (75-138 A.D.) put to death the architect Apollodurus of Damascus for having the temerity to criticize the plans for the Temple of Venus.[46]

But all in all, engineers in Rome, at least those engaged in large government-sponsored projects, enjoyed a high social status. Indeed, civil and military engineering was the only segment of the realm of technology that was not considered to be servile and degrading.[47] There seems to have been an awareness, whether conscious or not, that Rome's greatness rested on the skill of her engineers as much as on the strength of her legions and their commanding generals.[48] The legions were magnificently organized, but they would have been far less effective without the roads that they marched on and the engines of war that they used to such good effect. Certainly we know as much about Roman civilization from the structures and artifacts it left behind as from what has survived of the Roman literary record.

If engineers mold their civilization, they are also molded by it, and the Roman engineers were no exception. Their civil engineering, like the Roman imperium itself, was rigid and unyielding. On occasion, their structures could demonstrate an impressive level of engineering efficiency; the Pont du Gard, which carried the aforementioned Aqueduct of Nimes, was built with a safety factor of about two in regard to cracking— a margin similar to the engineering standards of today.[49] But many of their structures were needlessly heavy and over-designed, as were their roads, superior though they were to all other European roads prior the eighteenth century.

Part of the reason for this was technical; the Romans lacked instruments and techniques for accurate calculation of loads and stresses. But no less important, there was little incentive to develop these techniques at a time when the cost and supply of labor did not have to be considered, and local and relatively cheap materials were available for most purposes. We can also note the opposition of the emperor Vespasian (69-79 A.D.) to a water-powered construction hoist, on the ground that it would deprive the poor of work.[50] Fears of technological unemployment, groundless though they were, concerned the Romans as much as they have concerned later generations.

Rome at its height had profitable mining operations from Britain and Spain in the West to Asia Minor in the East. These, along with the elaborate water-supply systems, required well-developed pumping technologies. The Romans used the Archimedean screw and the noria, both of earlier origin. They also were aware of the principle of the force pump, although they were unable to use it effectively. They were much more successful in their development and use of the vertical water wheel, which became a significant source of power in the later days of the Empire, perhaps as a response to the labor shortages caused by the decline of slavery. The steady decrease in the size of the Empire's population that began in the third century A.D. was undoubtedly a contributing factor as well.

But on the whole, Roman engineers did little to develop mechanical devices except, as could be expected, in the military sphere. They did make more extensive use of such mechanisms as were already known, simply because the Roman Empire did things on a larger scale than anything that had gone before in Europe. The success of Roman engineering really owed more to the ability to organize large-scale projects than it did to unique technical capabilities.[51]

The Ancient Engineers in Perspective

Although some promising starts were made with new sources of power and labor-saving technologies, the ancient world was not a fertile ground for the invention and application of mechanical devices. This inability or unwillingness to employ mechanical devices runs through the whole of ancient civilizations, with no simple and completely satisfactory explanation of why this was so. It certainly was not due to lack of potential talent. We have already noted a number of examples of technical cleverness. To these can be added a Greek astronomical calendar that employed a highly complex arrangement of gears and linkages; as the scholar who made a careful examination of this device noted, "Men who could have built this could have built almost any mechanical device they wanted to."[52]

While the ancient engineers did not lack for ingenuity, many of their greatest intellectual achievements had no practical consequences. It is one thing to conceive of a technical innovation, and quite something else to execute it, especially when it is to be put to widespread use. Heron and Ctesibius could devise ingenious mechanisms and even make demonstration models, but they were in the same position that Leonardo da Vinci was to occupy later. They could visualize the way to go, but they had no way of making their ideas effective. Lacking a supportive substructure of materials, skills, and complementary devices, many of the their ideas could not be realized in practice.

To some extent, the slow development and application of mechanical technologies can be blamed on philosophical orientations. In the Hellenic world, the disdain for the practical employment of knowledge expressed by some philosophers had some effect, but even the utilitarian Romans did not depart from the Greeks in this regard. A more concrete reason may have been the pervasive influence of slavery. Forced servitude was widely used in the ancient world for public and private work. Alternatively, as in Egypt, public works were built by corvee labor, which could be equally coercive. Labor was plentiful, and once slaves were obtained there was little point in lightening their load. As a result, there were few reasons to create labor-saving devices. At the same time, the management and control of slave labor required that work be simple and routinized, resulting in the stagnation of the technologies employed.[53] Perhaps most importantly, the pervasive use of slavery produced a disdain for manual work, for it was seen as the special province of unfree men and women. Under these circumstances, the inventor of a production-increasing device or process occupied a region too close to the realm of a slave.

The major source of incentives for technological innovation was the military realm. In siege operations mechanical devices could do things that no amount of unaided human muscle could accomplish. Accordingly the Hellenistic and Roman periods accordingly saw the development of a number of ingenious siege weapons. Significantly, for many centuries the word "engine" referred only to catapults and similar weapons of war. We have seen how the technical genius of men like Agrippa and Archimedes was used for martial purposes, a pattern that has continued to our own times.

In practical terms, the engineers of the ancient world encountered many obstacles caused by a lack of adequate tools and materials. Cutting tools were made of bronze in the early years of recorded civilization and of iron of uncertain quality later. Steel was known, but the means of producing it relied on small-scale handicraft techniques, rendering it too expensive for any but highly specialized purposes. Measuring instruments were elementary. The *groma*, known from Egyptian times, was used for establishing right angles and straight lines for short distances. It was simply a horizontal cross with plumb lines at the end of each.[54] Heron's improved instrument, the *dioptra*, was little used, due to the difficulty of constructing it. The Romans also used a leveling instrument called the *chorbates*, which was simply a straightedge mounted on two equal legs, with a water trough on top to give greater accuracy. The mathematical tools of the modern engineer did not exist, and the systems of mathematical notation used in the ancient world made any elaborate calculation almost impossible. Accordingly, there was no way to calculate stresses or the strength of materials. This was one of the major reasons for the massive solidity of Roman construction; it was the only known way of providing a margin of safety.

Engineering also suffered from the slow transmission of useful ideas. Much of this was deliberate, as many engineers sought to keep their methods secret. At the same time, however, many things that might have been shared willingly remained bottled up. Frontinus was stymied by the inability to measure the flow of water in aqueducts he supervised, even though several decades earlier the engineer in charge of the Aqueduct of Nimes had devised an ingenious solution to the problem. But in the absence of instruments of communication such as handbooks, technical journals, and professional meetings, the technique apparently never travelled from Gaul to Rome.

Finally, it is important to note that the work and accomplishments of the ancient engineers were constrained by the necessity to adhere to the needs of their employers. Most engineers were in the employ of government authority of one kind or another, for the most part emperors and kings who wanted to glorify their reigns, hence the monumental projects typical of those times. Occasionally, however, the engineers got the better of their employers. During the third century B.C. the Lighthouse of Alexandria, one of the Seven Wonders of the Ancient World, was designed and built by Sostratos of Knidos. As was the custom, the ruler of Alexandria, Ptolemaios (or Ptolomy) wanted his name, and his name only, to appear on the structure. Instead, Sostratos had inscribed "Sostratos Son of Dexiphanes of Knidos on Behalf of All Mariners to the Savior Gods." This was covered with plaster, upon which a conventional inscription giving credit to Ptolemaios was carved. When the plaster eventually crumbled, the king's inscription fell away, revealing the one honoring Sostratos.[55]

In many parts of the ancient world it was also customary for wealthy individuals to fund substantial public works; in fact, it was expected of them. At the same time, however, the wealthy were markedly uninterested in sponsoring advances in productive technologies. Riches were not the product of industry, but of war and politics, and the most favored repository for investment was land, not capital goods.[56] All in all, the social and economic environment was receptive to only a limited range of engineering projects.

With the upper echelons of the society providing most of the patronage, it is not surprising that engineers were often expected to have the "right" social background. Vitruvius approvingly reported that Greek practice was to entrust civil engineering projects "in the first place to master builders of good family and next to inquire whether they had been properly educated— and the master builders themselves would teach none but their own sons or kinsmen, and trained them to be good men."[57]

This elevation of family status over proper education would not be accepted in a meritocratic society, but the very idea of technical qualifications counting for more than social origins is a very recent idea, and one that has hardly been fully realized even today. Under these circumstances, engineering could seldom be practiced as a free-standing profession; instead, it was enmeshed in a variety of social, governmental, or military roles. We will return to these themes in later chapters, but before we do, we need to consider the fate of engineers and engineering during the centuries following the collapse of the Roman Empire.

Notes

1. Norman A.F. Smith, "Attitudes to Roman Engineers and the Question of the Inverted Siphon," *History of Technology* 1 (1976), p. 49.

2. R.H. Forbes, *Studies in Ancient Technology*, vol. 1 (Leiden: E.J. Brill, 1955), pp. 156-57.

3. The quotation is from Byron, "The Destruction of Sennacherib." For a Biblical account, see 2 Kings 18, 19.

4. Forbes, op. cit., p. 158.

5. Richard Shelton Kirby, et al., *Engineering in History* (New York: McGraw-Hill, 1956), p. 32.

6. L. Sprague de Camp, *The Ancient Engineers* (Garden City, N.Y.: Doubleday, 1963), p. 44.

7. Ibid.

8. Reginald Englebach, *The Problem of the Obelisks: From a Study of the Unfinished Obelisk at Aswan* (London: T.F. Unwin, 1923), pp. 95-96.

9. James H. Breasted, *A History of the Ancient Egyptians* (New York: Charles Scribner's Sons, 1908), p. 159.

10. Forbes, op. cit., p. 23.

11. *Politics* 1278A, quoted in Alison Burford, *Craftsmen in Greek and Roman Society* (Ithaca: Cornell University Press, 1972), p. 34.

12. K.D. White, *Greek and Roman Technology* (Ithaca: Cornell University Press, 1984), p. 91.

13. Robert Scranton, "Greek Building," in Carl Roebuck (ed.), *The Muses at Work: Arts, Crafts, and Professions in Ancient Greece and Rome* (Cambridge, Massachusetts and London: MIT Press, 1969), pp. 8-10.

14. Ibid., pp. 11-12, 15.

15. de Camp, op. cit., p. 100. See also June Goodfield and Stephen Toulmin, "How Was the Tunnel of Eupalinus Aligned," B.L van der Waerden, "Eupalinos and His Tunnel," and Alfred Burns, "The Tunnel of Eupalinos and the Tunnel Problem of Hero of Alexandria" in Otto

Mayr (ed.) *Philosophers and Machines* (New York: Science History Publications, 1976).

16 Herodotus, *The History of Herodotus*, trans. George Rawlinson (New York: Tudor, 1928), p. 269.

17 Ibid., p. 270.

18 de Camp, op. cit, p. 147.

19 Plutarch's oft-quoted assessment of Archimedes appears in Friedrich Klemm, *A History of Western Technology* (Cambridge: MIT Press, 1964), p. 22.

20 Werner Soedel and V.L. Foley, "Archimedes: Engineer First and Mathematician Second", paper presented at the annual meeting of the American Association for Engineering Education (Knoxville, Tennessee, 15 June 1976)

21 This statement appears in Plutarch, *Lives, Marcellus*, xiv, 6-9; xv, 1-4; xvi, 2.

22 Portions of the account presented in A. Bouche-Leclerq, "L'Ingenieur Cleon" *Revue des Etudes Grecques*, 21 (1908) can be found in Finch, op. cit., pp. 36-38 and Forbes, op. cit., pp. 27-28.

23 Bouche-Leclerq, op. cit., p. 147.

24 Ibid., p. 147.

25 J.G. Landels, *Engineering in the Ancient World* (Berkeley: University of California Press, 1978), pp. 28-30.

26 Robert Temple, *The Genius of China* (New York: Simon and Schuster, 1989), pp. 215-48.

27 Joseph Needham, *Science and Civilization in China, vol. 4, part 3: Civil Engineering and Nautics* (Cambridge: At The University Press, 1971), p. 319.

28 "Hydraulic Engineer Li Bing," *China Reconstructs*, 38, 5 (May 1989) pp. 57-58.

29 Temple, op. cit., pp. 182-82.

30 Joseph Needham, *Science and Civilization in China, vol. 4, part 2: Mechanical Engineering* (Cambridge: At the University Press, 1965), p. 33.

31 Ibid., pp. 36-37.

32 James K. Finch, *The Story of Engineering* (Garden City, New York: Doubleday, 1960), p. 16.

33 White, op. cit., pp. 202-203.

34 Finch, op. cit., p. 75.

35 de Camp, op. cit., pp. 212-16.

36 Landels, op. cit., p. 213.

37 de Camp, op cit., p. 200.

38 For technical details, see George F.W. Hauck, "The Roman Aqueduct of Nimes" *Scientific American* 260, 3 (March 1989).

39 de Camp, op. cit., pp. 164-65.

40 Vitruvius, *Vitruvius on Architecture*, vol. II, trans. Frank Granger (Cambridge, Massachusetts: Harvard University Press, 1956) p. 7.

41 Ibid., p. 9.

42 Vitruvius, *Vitruvius on Architecture*, vol. 1, trans. Frank Granger (Cambridge, Massachusetts: Harvard University Press, 1956), p. 7.

43 Ibid., pp. 7, 9.

44 Vitruvius, op. cit., vol. 2, p. 7.

45 Vitruvius, op. cit., vol. 2, p. 271.

46 James E. Packer, "Roman Imperial Building 31 B.C.-A.D. 138," in Roebuck, op. cit., p. 41

47 Henry Hodges, *Technology in the Ancient World* (New York: Knopf, 1970), pp. 211-13.

48 It would be nice to report here that the honorific title given to the Chief Priest, *Pontifex Maximus* (from which is derived the Pope's title of *Pontiff*) means "master bridge-builder." Unfortunately, current scholarly opinion holds that the title is derived not from *pons* (bridge), but from the ancient word *puntes*, meaning religious rites and sacrifices). See Marjorie Nice Boyer, *Medieval French Bridges: A History* (Cambridge, Massachusetts: The Medieval Academy of America, 1976), p. 2

49 Hauck, op. cit., p. 102.

50 Charles Singer, et al., *A History of Technology*, vol. 2 (London: Oxford University Press, 1956), p. 601.

51 Donald R. Hill, *A History of Engineering in Classical and Medieval Times* (LaSalle, Illinois: Open Court, 1984), p. 52.

52 Derek J. de Solla Price, *Science Since Babylon* ed. 2 (New Haven: Yale University Press, 1975), p. 48.

53 Joel Mokyr, *Lever of Riches: Technological Creativity and Economic Progress* (New York and London: Oxford University Press, 1990), p. 195.

54 Kirby, et al., pp. 81-82.

55 de Camp, op. cit., pp. 125-26.

56 M.I. Finley, "Technological Innovation and Economic Progress in the Ancient World," *Economic History Review*, second series, 18, 1-3, (1965), p. 39.

57 Finch, op. cit., p. 38.

Chapter 3

The Medieval Engineer

Historical terminology does a considerable disservice to the period that stretched from approximately the sixth to the fifteenth centuries A.D. According to conventional periodization, the collapse of the Roman Empire in the West was followed by the Dark Ages, a period of social chaos, ignorance, superstition, and the near extinction of classical learning. The broader term, Middle Ages, suggests a kind of nondescript interval between the glories of the ancient world and the dynamic modern era ushered in by the Renaissance. And finally, the term "Gothic," which is used to describe much of the era's artistic creations, was initially synonymous with "barbaric."

In fact, the medieval era was a period of substantial progress, not the least of which was in the realm of technology. The "barbarians" who took over Europe may have known little of Greek art, philosophy, or science, and they had little of the Romans' administrative skills, but their practical abilities were most impressive. Over the course of the Middle Ages, developing technical skills were enhanced by the rediscovery of Greek and Roman texts, many of which had been preserved in the Muslim world, which itself became a source of new technical knowledge.[1] Further to the East, major technological innovations continued to appear in China, many of them diffusing to Europe. The blending of indigenous and imported artifacts and ideas produced a technologically dynamic society that in many ways surpassed the exalted civilizations of Antiquity, while at the same time paving the way for the even greater accomplishments that were to come.

The efforts and achievements of medieval engineers and workers found a ready outlet in a culture that manifested a growing enthusiasm for material progress and the subjugation of nature.[2] On a practical level, many new devices and sources of energy were eagerly accepted as substitutes for the large reserves of manpower typical of the ancient world. Although the early Christian church did not forbid slavery, it certainly disapproved of enslaving fellow Christians through warfare. Not that there were many opportunities to do so; for much of the early Middle Ages, Europe was on the defensive; as a result there were few opportunities for acquiring slaves through wars of conquest.[3] Still, one should not be too quick to attribute technical virtuosity to the stimulus of chronic labor shortages. A history of Chinese technological development shows that a vast labor force did not inhibit the creation of an impressive number of labor-saving inventions.[4]

Perhaps the most positive benefit that came from the withering of slavery was that it gave a new dignity to work. Although the assertion that work took on a spiritual dimension during the early years of Christianity has been questioned,[5] it nonetheless seems likely that even physical work had become a source of pride. Under these circumstances, there was little stigma attached to the labors of engineers who found it necessary to work with their hands as much as with their brains.

An Era of Innovation

In considering the sources of European technological progress, it is important to understand that not all inventions were home-grown. The technological successes of the Middle Ages owed much to a willingness to make effective use of other people's inventions. The Romans, for all their practical skills, never succeeded in as simple an endeavor as designing a horse harness that would not choke the unfortunate beast that strained against it. The adoption of an effective horse collar first used in China[6] made for considerable improvements in the speed at which fields could be plowed and other agricultural work done. Along with the horseshoe (possibly also imported from China), the improved harness directly led to the production of greater food reserves and a better-nourished population.

Medieval people also made good use of another foreign (either Chinese or Indian) invention, the stirrup. This seemingly obvious device transformed the nature of warfare by making the mounted knight the decisive agent of war. At the same time, the pre-eminence of the knight solidified the political-economic system that we now call feudalism, for an elite fighting force had to be supported by an agricultural economy based on the labors of serfs, who in turn looked to the knight for defense in troubled times.[7] Knightly combat was the basis of warfare until the 14th century when English longbowmen and Swiss pikemen showed that the armored knight was not always invincible. At about this time the knight's fate was further sealed by the use of firearms, a medieval innovation that combined indigenous inventive ability with another import from China, gunpowder.

While improved means of waging war may have provided no overall benefits for Medieval people, their lives were considerably enhanced by new technologies for producing the most direct source of energy: food. In addition to the much more effective use of horse power, European cultivators increased the yield of their fields by adopting the three-field system of crop rotation. Instead of leaving half of their fields fallow in order to restore the fertility of the soil, farmers left only one-third of their fields fallow, while another third was devoted to the growing of grain and the remaining third was given over to the cultivation of oats, barley, peas, lentils, and broad beans. Oats were particularly valuable as food for horses, while the last three crops had the double benefit of fixing more nitrogen in the soil while at the same time providing a source of needed protein.[8]

Medieval ingenuity was not confined to agriculture and warfare. Foremost among the technological accomplishments of the era was the effective use of inorganic sources of power. In the words of Lynn White, "The chief glory of the Middle Ages was not... its cathedrals, its epics, its vast structures of scholastic philosophy, or even its superb music; it was the building for the first time in history of a complex civilization which was upheld not on the sinews of sweating slaves and coolies but primarily by non-human power."[9]

Although the water wheel was known to the ancient world for the milling of grain, during the Middle Ages it was put to a vast number of new uses: pumping water out of mines, powering bellows for iron smelting, sawing wood, grinding and polishing weapons, preparing mash for beer-making, and running trip hammers for forging iron, pulping paper, grinding paint pigments, fulling cloth, and extracting tannin to tan leather.[10] In 1086 the Domesday Book recorded over 5000 water mills of various kinds throughout England. Most of them must have been quite small, for the above figure implies a ratio of one mill for every fifty households,[11] but what the water wheel lacked in size it made up in extensive deployment. The crank, unknown in ancient times anywhere but in China, was used in Europe by the 9th century A.D. The spinning wheel, another example of rotary motion, successfully diffused from China. China may also have been the source of the sternpost rudder, which, along with the lateen sail and astrolabe, allowed ships to sail into the wind and navigate with greater precision

It would be most helpful to know something about the engineers and craftsmen that designed and built these devices. Unfortunately, the historical record is almost entirely silent on this count. Although many occupied the role of engineer, they would not have been called by this title, for at this time the term "ingeniator" was almost exclusively reserved for the designers of military equipment, especially catapults and other siege weapons.[12] But engineers they were, and though many of their technical skills may have been inferior to those of today's engineers, their accomplishments are among engineering's greatest triumphs.

The Master Mason and His Work

Fortunately, we know at least a little about a few people who could be classified as civil engineers. The political fragmentation typical of the Middle Ages prevented the construction of works on the scale of the Great Wall and Grand Canal of China, or the roads and aqueducts of Rome. Still, many imposing structures were constructed, culminating with the soaring vaults and spires of the Gothic cathedral.

Long before this occurred, the greatest example of church architecture was completed in Constantinople. While the Roman empire in the West was going through its death throes, the Byzantine empire in the East flourished, with the church of Holy Wisdom (Hagia Sophia) standing as its most impressive monument to engineering. Little is known about the sixth-century architect-engineers who built this great edifice, Anthemios of Tralles and Isidoros of Miletus. They were both mathematicians as well as architect-engineers, and Anthemius was noted for his studies of optics and mechanics.[13] There is also a story that he created a simulated earthquake through the use of steam pressure in order to frighten an annoying neighbor.[14]

There are also a few snippets of information about the engineers responsible for two famous bridges, both immortalized in song. The Pont d'Avignon was begun in 1177 by the man who would be canonized as St. Benezet. He died before the completion of the bridge ten years later, and was buried in a chapel in one of its piers. St. Benezet was a shepard boy who became a member of a monastic order known as the Freres Pontifes.[15] According to legend, he was commanded in a vision to build the bridge and to instruct others how to do it, his credibility greatly enhanced by his miraculous moving of a stone that had resisted the efforts of thirty men. London Bridge was begun just a year earlier (1176) but its story is more prosaic. Its designer was Peter of Colechurch, also a priest. He died in 1205, four years before his bridge was finished, and like St. Benezet was buried in its chapel.[16] His work was completed by three London merchants who engaged the services of a Brother Isembart, a French pupil of Benezet. Peter and Isembart built a bridge that survived for 600 years until it was replaced early in the 19th century by the bridge designed by John Rennie and his son, a bridge that had the odd fate of eventually being transported to the Arizona desert where it now spans an artificial lake.

The reappearance of the priest-engineer is understandable, for through much of the Middle Ages literacy was pretty much restricted to the clergy. Even so, the builders of the Gothic cathedral, the greatest technological accomplishment of the era and its preeminent symbol, were generally not clergymen. Rather, they were referred to as master masons. In fact, they were far more than the name implies, for they combined a great multiplicity of roles. As John Fitchen summarizes, in late medieval times there was

> neither the specialization nor the separation of functions that is reflected in our design and operation personnel today: the architect, the structural engineer, the various mechanical engineers, and the general contractor, together with his team of sub-contractors. The medieval master builder was really a master of all phases of the work, familiar with each operation and constantly in immediate touch with it. Hence, he was an imaginative and creative designer... who had to be comprehensively and intimately familiar, at the same time, with the means by which the design could be brought to realization in actual stone and mortar.[17]

One example of a multi-talented master mason was William of Sens, the builder of Canterbury Cathedral. The admiring monks who documented his efforts noted that he designed "the most ingenious machines for loading and unloading ships, and for drawing the mortar and stones." The design of these machines must have called for a high level of engineering skill; the hoists he used were probably capable of lifting about half a ton.[18]

The master masons of the Middle Ages were the successors of the Roman *architecturus*, with much the same relationship to their employers. Their products of their skills are impressive even today. The vaults of Gothic cathedrals soared to breathtaking heights; a fourteen-story building could be built inside Beauvais cathedral and still not reach the vaulting. The spire of the cathedral at Chartres is equivalent to a thirty-story building, and that of Strasbourg stretches as high as a forty-story skyscraper.[19] A great deal of clever structural engineering went into their design. Research using modern methods of stress analysis shows the essential correctness of typically Gothic elements such as vault webbing, the flying buttress, and the placement of heavy pinnacles atop pier buttresses.[20]

The master mason was charged with both the design of the cathedral and its actual realization in stone and mortar. His skills were anything but commonplace, and municipalities often had to hire master masons from far away to work on their cathedrals. Villard Honnecourt, about whom more will be said shortly, travelled all the way from his native France to Hungary to oversee one project, and another French master mason took his skills to Sweden, at that time a far frontier of Western Christendom.

Their services did not come cheap. One record of the building of Carnarvon castle in the fourteenth century shows that the master mason received two shillings each day, while ordinary masons received two shillings sixpence each week.[21] It can also be noted that John of Gloucester, the king's mason, owned four houses and a country estate, was paid double salary when traveling on business, and was presented with two squirrel robes annually, as well as two more for his wife. Despite this largesse, John died owing the king the sum of £ 53 6s. 8d. James of St. George, who was responsible for the building of a chain of fortresses in Wales during the late thirteenth century, earned a salary in cash and other benefits of £ 80 a year, this at a time when the ownership of £ 20 worth of land entitled the holder to a knightship.[22] Many master masons further enriched themselves by acquiring the quarries that produced the stones used in cathedral construction,[23] a potential conflict of interest that does not seem to have troubled medieval people.

The high status of medieval master masons was not reflected in just their salaries. Master masons were often honored by prominent inscriptions within the cathedral that called attention to their builder, as exemplified by the 25-foot long inscription at the base of the South transept of Notre Dame de Paris: "Master Jean de Chelles commenced this work for the glory of the Mother of Christ on the second of the Ides of the Month of February, 1258."[24] Many master masons were buried in the churches they had built, their final resting place marked by inscriptions of which the following is typical: "Weep, for here lies buried Pierre, born at Montreuil, a pattern of character and in his life a Doctor of Masons. May the King of Heaven lead him to the skies: he died in the year of Christ, the thousandth, two hundredth, with twelve and fifty four."[25]

Even more significant were the centrally located labyrinths outlined in stone on the floor of the cathedral nave. These mazes represented a pilgrimage to Jerusalem, and many of the faithful would crawl along them while on their knees in order to attain spiritual benefits. The maze was followed to the center, in which rested a marble or metal plaque that depicted not some holy figure, but the master masons responsible for the cathedral's construction. Their portraits were accompanied by their names, dates, and the roles that they had played in the cathedral's construction.[26] The cathedral at Reims portrayed four of its builders at the corners of the maze. Two were depicted with the tools of their trade: one with a square rule, and the other wielding a large compass with which he was laying out a rose window.[27] By the thirteenth century the master mason could even serve as a model for the Creator, with the Almighty depicted using a compass to measure the world He had created.

The exalted status of master masons was evidently a source of annoyance for some. One thirteenth-century clergyman seems to have been bothered by the fact that craftsmen had risen to the rank of supervisors, although his real target is the church hierarchy, whose attitudes and manner bore a strong resemblance to those of the master masons:[28]

> The master masons, holding measuring rod and gloves in their hands, say to others; "Cut here," and they do not work; nevertheless they receive the greater fees, as do many modern churchmen.
> Some work with words only. Observe: in these large buildings there is wont to be one chief master who orders matters only by word, rarely or never putting his hand to the task, but nevertheless receiving higher wages than the others. So there are many in the church who have rich benefices, and God knows how much good they do; they work with the tongue alone, saying "Thus should you do," and they themselves do nothing.

It can be reasonably expected that master masons were proud of their skills, and had a strong sense of their individuality. Certainly the notion that the designers and builders of medieval cathedrals were anonymous, humble workmen is little more than a charming myth. Still, there were limits on the master mason's independence. It was not unusual to call in other master masons to inspect and approve the plans of the master mason responsible for a project.[29] There is an early fifteenth-century record of one church chapter which had some doubts about the structural design that had been submitted for the completion of their cathedral. They therefore called in eleven other master masons from other church projects, who agreed that the plan was a safe one.[30]

Master masons viewed themselves as an elite body, and from early on they had come to a collective understanding not to disclose their technical knowledge to those outside their own circle.[31] Still, it was not until quite late in the Middle Ages that they developed well-organized guilds to preserve and monopolize technical craft secrets.[32] In the end, masons' guilds became restrictive, resistant to invention, and almost like castes, but for much of the period they fulfilled a positive function. The guilds provided an intellectual support structure for the training of new masons, as well as being repositories of technical knowledge and skill.

The Design Techniques of the Medieval Engineer

Despite efforts by the medieval guilds to restrict the spread of knowledge outside their ranks, a great deal of technical information was disseminated through the community of medieval engineers. Cathedral construction was slow, often agonizingly so, giving plenty of time for techniques to be transferred to new practitioners. The trans-European scope of cathedral building also resulted in a great deal of technology transfer among architect-engineers. Construction sites have been aptly described as "genuine permanent schools" from which was transferred a vast knowledge based on the direct observation of successes and failures.[34] Here can be found a clear example of how the historical diffusion of technology is in large measure the result of the migrations of skilled personnel.[35]

Master masons learned their craft directly from other masons, and not by reading abstract discussions of the theory and practice of vault construction.[36] It is likely that many engineers of the early Middle Ages were not literate, relying instead on clerks to keep business and construction records.[37] The limited importance of printed sources of technical knowledge is indicated by the complete absence of extended treatises on construction during the great cathedral-building era.[38] Still, there were elements of formal instruction, and as literacy became more prevalent during the late Middle Ages, it is likely that aspiring master masons had to pass written examinations that tested their skills as designers.[39]

The knowledge base of the master mason was "a remarkably well-organized empiricism,"[40] augmented by a sophisticated application of practical geometry. This was an essential skill, for in the absence of the ability to calculate forces and stresses, correct design rested on getting the shape right, and that meant using geometry properly.[41] In producing their designs, master masons used little more than a straightedge, compass, square, and dividers. The basic proportions of a building were expressed in modular units. Plans were employed, but they rarely provided actual units of measurement such as feet and inches.[42] This was not a serious limitation, for plans were generally not used as working drawings; rather, they were used to impress patrons, raise funds for construction, and serve as aids in choosing between design alternatives.[43] The drawings prepared by the master mason left out many details, and did not specify the size of stone blocks and how they were to be fitted together.[44] In order to translate plans into reality, it was usually necessary to draw full-sized renderings of the plans onto the plaster floor of a tracing house located on the building site. These drawings then were used for the construction of wooden templates that in turn guided masons as they shaped stones into the required components of the building.[45]

Mathematical calculations were rarely done; even simple multiplication was beyond the abilities of most master masons.[46] Yet this was not a serious hindrance. Through the use of simple drafting tools and the application of step-by-step methods, master masons could perform such essential tasks as deriving the elevation from a ground plan or determining the circumference of a circle.[47] These procedures allowed them to produce marvelously sophisticated designs, such as elegant rose windows based on complex patterns of inscribed squares and circles.[48]

The completion of design work did not mark the end of the master mason's task; it was still necessary for him to directly supervise the actual construction work. This must have been a herculean task; one scholar has noted that a particular project entailed the supervision of a "labour force of 400 masons, 2000 minor workmen, 200 quarrymen and 30 smiths and carpenters, together with a supply organization of 100 cars, 60 wagons and 30 boats bringing stone and sea-coal to the site."[49] The supervision of construction by the master mason gave him an intimate familiarity with every facet of building design and construction.[50] His involvement with all aspects of the work gave a distinctive stamp to this branch of engineering during the Middle Ages. Master masons had an intimate knowledge of the materials they used and how the construction proceeded. The result was a combination of form and function that has rarely been surpassed.

The Ideas and Accomplishments of One Medieval Architect-Engineer

The wide range of interests and skills possessed by medieval master masons is illustrated by the work of Villard de Honnecourt. We know something about his interests and accomplishments because 33 leaves of his notebook survived the ravages of time before being discovered toward the end of the eighteenth century.[51] Of Villard himself we do not know a great deal. He was born in Picardy in Northern France at the turn of the thirteenth century and plied his trade as a master mason between 1225 and 1250. While still a boy he drew up the ground plan for the choir of the church of a Cistercian abbey near his home. He went on to work on a number of other church construction projects, including one, as was noted earlier, in Hungary. His notebook is full of designs extracted from Vitruvius and Hero of Alexandria, as well as sketches of things he saw and things that he did or intended to do.

Villard's interests and activities qualify him as a Renaissance man who lived more than two hundred years before the Renaissance. But above all he was an engineer, even if this is not the title that would have been used in his lifetime. Along with material relating to building design and construction, his notebook contains references to several mechanisms, some of which may have represented Villard's own ideas. One sketch shows a sawing device for cutting pilings flush under water so that a platform could be built on them. One of his most frequently reproduced drawings shows a saw driven by a water wheel that uses a spring pole to provide the return stroke. It also incorporates a means of mechanically feeding the log into the saw, thereby imparting a measure of automatic operation for the whole process. Villard also sketched clocks and escapements, and more quixotically gave some suggestions for a perpetual motion machine using a wheel that was to be driven by the falling of mallets loaded with quicksilver.[52] We now know that perpetual-motion machines are impossible, even absurd, but Villard's interest in them shows a keen interest in mechanical sources of power, an interest that marked the beginnings of a new era of European thought and action. As Lynn White has argued, medieval people "were coming to think of the cosmos as a vast reservoir of energies to be tapped and used according to human intentions. They were power conscious to the point of fantasy. But without such fantasy, such soaring imagination, the power technologies of the Western world would not have been developed."[53]

Aeronautical Engineering in the Middle Ages

A different kind of fantasy gripped an early eleventh-century astronomer and mathematician named Eilmer of Malmsbury. Eilmer hoped to fly like a bird, and to that end he constructed some sort of glider.[54] It appears to have consisted of rigid wings attached to his arms and legs and probably intended to flap. Eilmer jumped out of a tower and flew for about 600 feet until the instability of his craft resulted in a crash landing and two broken legs. He attributed the unfortunate termination of his experiment to his failure to put a tail on his glider. It is significant that Eilmer had a matter-of-fact explanation for what had happened. Neither he nor the chronicler of his flight attributed his crash to supernatural intervention. He flew, he fell, and was crippled so that he never could try the experiment again, and all that he said was that he should have remembered that birds land tail down. In the conventional view of the Middle Ages there should have been some reference to evil spirits or Divine disapproval ("If God wanted men to fly..."). What Eilmer of Malmsbury has to tell us is that these times were not entirely buried in ignorance and superstition, and that there must have been considerable cultural support for empirically based engineering.

Similar but more successful efforts at aeronautical engineering were made in China, although unlike the more prosaic horsecollar, these innovations never diffused to Europe. The first aeronautical experiments had a cruel intention; they were used by the Emperor Gao Yang (r. 550-559) to combine entertainment with the execution of his enemies. Members of rival families were harnessed to large paper kites and thrown from high towers. In one recorded instance, a condemned prince was said to have made a successful flight of two miles, only to be subsequently imprisoned and starved to death. It is possible that Taoist priests subsequently engaged in kite-borne flights, although there is no hard evidence of it. In the 13th century Marco Polo recorded the use of man-lifting kites used for divination, although it must be said that the aviators were viewed as "fools or drunkards, for no one in his right mind or with his wits about him would expose himself to that peril."[55]

The Summit of Medieval Mechanical Engineering: The Clock

Chinese inventiveness was not limited to the accomplishments narrated in this and the previous chapter. China also took a leading role in the development of one of the most important devices ever invented: the mechanical clock. As Lewis Mumford has argued, a clock is the very model of a mechanical device: it makes use of a power source that provides an even flow of energy, thereby making possible automatic action and the regular production of a standardized product (in this case hours and minutes).[56] Its social consequences have been vast; to mention but one, the coordination of individual activities in a factory or office would be virtually impossible without the artificial divisions of the day served up by the clock.

The greatest engineering challenge in the building of early mechanical clocks was the design of an escapement mechanism to evenly regulate the clock's source of power. A falling weight will of course accelerate, making it necessary to convert its motion into one of uniform velocity. The motivation for designing a successful clock and escapement was rather curious, at least from a 20th-century engineering perspective. The Chinese imperial court was strongly imbued with a belief in astrology, and a precise calculation of time was necessary to determine the propitiousness of heavenly conjunctions when the emperor had a mind to impregnate the empress or a concubine with a potential heir to the throne.

In order to make this determination, an eighth-century Buddhist monk and mathematician named Yi Xing designed a water-powered astronomical instrument that also acted as a clock. It did not use a mechanical escapement; that function was served by having a water clock supply water to the paddles that drove the main wheel. Each filling of a paddle would turn the wheel after overcoming the resistance of a restraining pawl. Although the clock seemed to work well enough, the corrosion of its metal parts resulted in its being taken out of service. An improved version using mercury instead of water was built toward the end of the tenth century.[57]

The supreme achievement of medieval Chinese clock building was organized by Su Song at the end of the eleventh century. It too used water as both a source of power and as a means of regulation. Joseph Needham describes the mechanism thusly:[58]

> The wheel was checked by an escapement consisting of a sort of weighbridge which prevented the fall of a scoop until full, and a trip lever and parallel linkage system which arrested the forward motion of the wheel at a further point and allowed it to settle back and bring the next scoop into position on the weigh-bridge. One must imagine this great structure going off at full-cock every quarter of an hour with a great sound of creaking and splashing, clanging and ringing; it must have been very impressive...

Su Song was a diplomat and administrator who also had written extensively about pharmaceutical botany, minerology and zoology. The designer and superintendent of construction was Han Gonglian, a minor official in the Ministry of Personnel.[59] The prominent role played by government officials in a technical endeavor is not surprising, for the highly organized Chinese state had long siphoned off the best and brightest people, much as the Medieval Church had done in Europe. Sadly, the political connections of Su Song proved to be the clock's undoing, for it was eventually destroyed as a result of political infighting between Su and rival political factions.

No exact date can be set for the appearance of the first mechanical clock in Europe, nor can a single person be given credit as its inventor. About all that can be said is that mechanical clocks began to appear toward the end of the 13th century. Within a few decades clocks exhibited a high level of sophistication, as exemplified by the clock of Giovanni di' Dondi, which served as a miniature planetarium as well as a timepiece, and the great clock of Strasbourg cathedral that combined these functions along with "a pageant of moving figures, historical, mythical, and symbolic."[60]

The key technological breakthrough was the invention of the verge-and-foliot escapement, which used an oscillating movement to regulate the motion of a falling weight. One of the earliest clocks on record was built in 1335 by a monk with the financial backing of his superior.[61] Monasteries were natural sites for the construction of early clocks, for they had long relied on some means of timekeeping (such as sundials and water clocks) for the scheduling of prayers and other activities. Monasteries were also repositories of engineering talent. As the archbishop of Mainz said of one group of monks in 1248, "I have found men after my own heart. Not only do they give witness of unblemished religion and a holy life, but also they are very active and skilled in building roads, in raising aqueducts, in draining swamps... and generally in the mechanic arts."[62] It is therefore likely that many clockmakers came from monasteries with mining and metallurgical enterprises, and populated by men with a good grasp of mechanisms and the utilization of power sources.[63]

Despite the importance of monastaries as centers of technological innovation, many of the early clock engineers were drawn from the laity; blacksmiths and other metal workers were especially evident, as were engineers and craftsmen who had experience with water mill design and construction.[64] Other skilled trades were represented; one early clock designer had been previously employed as the pope's plumber.[65] But practical skills did not always perfectly complement technical skills. It could not be expected that all blacksmiths would be engineering virtuosos, just as good engineers might not be very proficient at metal working. As a general rule, the designers were better at their trade than the blacksmiths were at theirs. The result often was, in Abbot Usher's words, "A strange combination of brilliance in conception with a deficient technique of construction."[66]

This does not mean that the designer's skills could be taken for granted. Clockmakers capable of engineering a good timepiece were scarce and much in demand. The clocks that adorned the townhalls and church towers of medieval Europe were expensive projects, costing the equivalent of perhaps a million dollars in today's currency. The master clockmaker got a good portion of that sum, and clockmakers in general seem to have received a comfortable income, even in comparison with skilled craftsmen. When three skilled workers were hired to repair and refurbish the clock adorning Cambrai cathedral they were paid less than two livres, while the master clockmaker received more than eight times as much.[67] Men skilled in clock engineering could even receive a tidy income for doing consulting work; one record shows that a smith brought in to provide advice on the repair of a cathedral clock received five livres, seventeen sous, and six deniers for his efforts.[68]

The Military Engineer

Sadly, not all of the medieval engineer's triumphs were in such peaceful realms as cathedral building and clock making. As will be recalled, the term *ingeniator* referred primarily to the designer of armaments. The product of his craft, the "engine," was a siege weapon such as a catapult or trebuchet. Medieval engineers also concerned themselves with a smaller weapon, the crossbow—another device that may have originated in China. So terrifying was this weapon that in 1139 the pope and the Lateran Council unsuccessfully attempted to ban its use against Christians. The trebuchet, which threw a projectile by means of a counterweighted lever, was one of the most useful weapons of medieval warfare. Additionally, it demonstrated the medieval interest in employing non-animate sources of energy—in this case gravity. The engineers who designed and built it occupied positions of considerable prestige and importance, as can be seen in the Count of Hainault begging "through love" his "very dear master of the engines" to make as big a machine as possible.[69]

Like clockmakers and master masons, military engineers were very much in demand throughout Europe. Many were quite happy to sell their services to the highest bidder, and in so doing they often travelled all over Europe. The autobiographical epitaph of Konrad Kyeser, a fifteenth-century military engineer, describes him as "A follower welcomed to their palace by many princes" and then duly lists seven rulers for whom he worked, followed by more than twenty cities, countries, and regions—everywhere from Sicily to Sweden, from France to Russia. With little modesty but perhaps some accuracy he claims that "His experience hath gained him a mastery of the art of war equalled by no man."[70] His book on military technology encompassed guns and gunpowder, battering rams, water pumps, lifting gear, vehicles of war, and even hot air balloons. If he truly were expert in these diverse fields he surely would have been much in demand, for there was no lack of combatants during the troubled times that he lived. The occupation of military engineer gave Kyeser a measure of wealth, fame, and prestige, but perhaps he wondered if it all had been worth it, for his epitaph laments that "There perished in a strange land Konrad Kyeser of Eichstatt... [who] now hath horrible death, most horrible, cast him down to Wrath. Thus lieth he bound, far from home, confined deep down in his coffin."[71]

Medieval Engineering in Transition

The medieval world contained no less technical talent than the ancient world, but it differed in the way it put that talent to use. In going about their work, medieval engineers largely relied on their own experience and cookbook procedures for producing the desired geometrical constructions. Abstract speculation was largely confined to the realm of theology, and the medieval world produced no equivalents of Archimedes. Yet towards the end of the Middle Ages technical applications began to be complemented by a more abstract mode of thinking. This tendency was at least in part due to increasing levels of literacy. As book-based learning began to supplement and even supplant on-the-job experience, the days of the illiterate engineer were numbered. As one early fifteenth-century textbook on guns and fireworks admonished, "The master must be able to read and write, for it is not possible that he retain in his mind all matters pertaining to his art..."[72]

The new technologies themselves stimulated abstract inquiry. Improved siege weapons kindled an interest in ballistics, the use of gunpowder stimulated chemical research, and improvements in shipbuilding made imperative a better understanding of astronomically based navigation. Engineering needs were reflected in mathematical inquiry, as when problems with the design of gear teeth for clockwork mechanisms prompted the mathematical analysis of the cycloid curve.[73] The result was not the complete triumph of scientifically based engineering, but these efforts did represent a step in that direction. This trend became even more evident during the next period to be considered, the Renaissance.

Notes

1 See Ahmad Y. al-Hassan and Donald Hill, *Islamic Technology: An Illustrated History* (Cambridge: Cambridge University Press, 1986)

2 The culture of the medieval period and the motivations of its inhabitants remain elusive, with much controversy surrounding efforts to characterize them. For an excellent summary of these issues, see Elspeth Whitney, *Paradise Restored: The Mechanical Arts from Antiquity through the Thirteenth Century* (Philadelphia: The American Philosophical Society), pp. 1-21.

3 Marc Bloch, *Land and Work in Medieval Europe* (New York: Harper & Row, 1969), pp. 181-82.

4 Joseph Needham, *Science and Civilization in China, vol. 4, part 2: Mechanical Engineering* (Cambridge: At the University Press, 1965), pp. 28ff.

5 See George Ovitt, Jr., *The Restoration of Perfection: Labor and Technology in Medieval Cultures* (New Brunswick and London: Rutgers University Press, 1987)

6 Robert Temple, *The Genius of China* (New York: Simon and Schuster, 1989), pp. 20-23.

7 This thesis has not gone unchallenged. For critical views, see R.H. Hilton and P.H. Sawyer, "Technological Determinism: The Stirrup and the Plow," *Past and Present*, 24 (1963) and Bernard S. Bachrach, "Charles Martel, Mounted Shock Combat, the Stirrup and Feudalism," *Studies in Medieval and Renaissance History*, 7 (1970)

8 Lynn White, Jr., *Medieval Technology and Social Change* (New York: Oxford University Press, 1962), pp. 69-76.

9 Lynn White, Jr., *Machina ex Deo: Essays on the Dynamism of Western Culture* (Cambridge, Massachusetts: MIT Press, 1966), pp. 70-71.

10 White, *Medieval Technology*, op. cit., p. 89.

11 Terry Reynolds, *Stronger than a Hundred Men: A History of the Vertical Water Wheel* (Baltimore: Johns Hopkins University Press, 1983), p. 52.

12 W.H. Armytage, *A Social History of Engineering* (Boulder: Westview, 1976), p. 45.

13 Richard Shelton Kirby, et al., *Engineering in History* (New York: McGraw-Hill, 1956), pp. 97-98.

14 White, *Medieval Technology*, op. cit., pp. 89-90.

15 Kirby, et al., op. cit., p. 108.

16 Ibid., p. 110.

17 John Fitchen, *The Construction of Gothic Cathedrals: A Study of Medieval Vault Erection* (Oxford: At the Clarendon Press, 1961), pp. xi-xii.

18 Arnold Pacey, *The Maze of Ingenuity: Ideas and Idealism in the Development of Technology* (Cambridge, Massachusetts: MIT Press, 1976), p. 32.

19 Jean Gimpel, *The Cathedral Builders*, trans. Teresa Waugh (New York: Grove, 1983), p. 7.

20 Robert Mark, *Experiments in Gothic Structure* (Cambridge, Massachusetts: MIT Press, 1982).

21 Gwilym Peredur Jones, "Building in Stone in Medieval Western Europe," in M. Postan and E.E. Rich (eds.), *The Cambridge Economic History of Europe, vol. 2: Trade and Industry in the Middle Ages* (Cambridge: Cambridge University Press, 1952), p. 507.

22 Jean Gimpel, *The Medieval Machine: The Industrial Revolution of the Middle Ages* (Harmondsworth: Penguin, 1977), p. 115.

23 Gimpel, *The Cathedral Builders*, op. cit., p. 97.

24 Gimpel, *The Medieval Machine*, op. cit., pp. 117-118.

25 Ibid., p. 119.

26 Ibid., p. 118.

27 Ibid.

28 Teresa J. Frisch, *Gothic Art, 1140-c.1450: Sources and Documents* (Englewood Cliffs, New Jersey, Prentice-Hall, 1971), p. 55.

29 Gerald A.J. Hodgett, *A Social and Economic History of Medieval Europe* (London: Methuen, 1972), p. 130.

30 Jones, op. cit., p. 497.

31 Gimpel, *The Cathedral Builders*, op. cit., p. 86.

32 Lon R. Shelby, "The Role of the Master Mason in Medieval English Building," *Speculum*, 39, 3 (July 1964), p. 46.

33 Bertrand Gille, "The Medieval Age of the West," in Maurice Daumas (ed.), *A History of Technology and Invention*, vol. 1: *The Origins of Technological Civilization*, trans. Eileen B. Hennessy (New York: Crown, 1969), p. 564.

34 Ibid., p. 537.

35 Carlo Cipolla, *Before the Industrial Revolution* (New York: W.W. Norton, 1976), p. 176.

36 Lon R. Shelby, "The Education of the Medieval Master Mason," *Medieval Studies*, 32 (1970)

37 Lon R. Shelby, "The Geometrical Knowledge of Medieval Master Masons," *Speculum*, 47, 3 (July 1972), p. 397.

38 A.C. Crombie, *Medieval and Early Modern Science*, vol. 1 (Garden City, New York: Doubleday, 1959), p. 205.

39 Francois Bucher, "Design in Gothic Architecture: A Preliminary Assessment," *Journal of the Society of Architectural Historians*, 27, 1 (March 1968), p. 57.

40 Gille, op. cit., p. 536.

41 Pacey, op. cit., p. 48.

42 Bucher, op. cit., p. 51.

43 Ibid.

44 Shelby, "The Role of the Master Mason," op. cit., p. 391.

45 Gimpel, *The Cathedral Builders*, op. cit., p. 94.

46 Mark, op. cit., p. 3.

47 Lon R. Shelby, *Gothic Design Techniques: The Fifteenth Century Design Booklets of Mathes Roriczer and Hanns Schmuttmeyer* (Carbondale: Southern Illinois Press 1977), pp. 64-66.

48 Bucher, op. cit., pp. 52-53.

49 A.J. Taylor, "Master James of St. George," *English Historical Review*, 65 (1950), p. 448.

50 Shelby, "The Education of English Master Masons," op. cit.

51 James K. Finch, *The Story of Engineering* (Garden City, New York: Doubleday, 1960), pp. 86-89

52 Gimpel, *The Medieval Machine*, op. cit., pp. 127-29.

53. White, *Medieval Technology*, op. cit., p. 134.

54. Lynn White, Jr., "Eilmer of Malmsbury: An 11th Century Aviator," *Technology and Culture*, 2, 2 (1961), pp. 97-111.

55. Temple, op. cit., pp. 175-79.

56. Lewis Mumford, *Technics and Civilization* (New York: Harcourt, Brace, and World, 1934), p. 15.

57. Temple, op. cit., pp. 105-107.

58. Quoted in Ibid., p. 109. For an extended discussion of this clock, see Needham, op. cit., pp. 450-65.

59. David S. Landes, *Revolution in Time: Clocks and the Making of the Modern World* (Cambridge, Massachusetts: Harvard University Press, 1983), pp. 17 and 393.

60. Ibid., p. 83.

61. William I. Milham, *Time and Timekeepers* (New York: Macmillan, 1923), p. 70.

62. Lynn White, Jr., "The Evolution of Technology, 500-1500," in Carlo Cipolla (ed.), *The Fontana Economic History of Europe: The Middle Ages* (Glasgow: Collins/Fontana, 1972), p. 170.

63. Landes, op. cit., p. 191.

64. Ibid.

65. Ibid., p. 194.

66. Abbot Payson Usher, *A History of Mechanical Inventions* (Cambridge, Massachusetts: Harvard University Press, 1954), p. 207.

67. Landes, op. cit., p. 200.

68. Ibid.

69. Phillippe Contamine, *War in the Middle Ages* (Oxford: Basil Blackwell, 1984), p. 194.

70. Friedrich Klemm, *A History of Western Technology* (Cambridge, Massachusetts: MIT Press, 1964), pp. 100-101.

71. Ibid., p. 122.

72. Ibid., p. 105.

73. Aubrey F. Burstall, *A History of Mechanical Engineering* (Cambridge, Massachusetts: MIT Press, 1965), p. 132.

Chapter 4

The Renaissance: The Recognition of the Engineer

Students of history used to be taught that the Renaissance began in 1453, when the capture of Constantinople by the Turks sent fugitive scholars to the West. Perhaps this is still taught; historical myths die hard. In fact, no single date or event can be given for the beginning of the Renaissance; the transition from the Middle Ages was gradual and almost imperceptible. But at least its birthplace can be located; the changing intellectual currents to which the term Renaissance is applied were first evident in Italy by the latter part of the 14th century, strife-torn and plague-ridden as it was. They then spread to the rest of Western Europe over the next two to three centuries.

It was during the Renaissance that the term "engineer" began to receive widespread currency. Engineering and architecture continued to be practiced largely by the same individuals, but there was a growing recognition of engineering as a separate activity requiring a distinctive set of skills. It was also a time when practicing engineers began to take a more individualistic and even aggressive stance: "In line with the Renaissance tendency towards uninhibited self-assertion, they promoted themselves, grasped for personal fame, and told off their rivals and employers when they thought themselves wronged."[1] One of the consequences of this new attitude is that in comparison with earlier times we know quite a bit more about the lives of men who were known as engineers, and this chapter therefore provides more information on the life and work of individual engineers than the previous chapters.

The most fertile ground for the cultivation of an engineer was 15th and 16th century Italy. As one French scholar asserted in 1567, Italy was the place to study or work "if anyone is a engineer [*mechanicus*] or an architect."[2] This in itself is a historical puzzle; there is no really satisfactory explanation for the flourishing of engineering in this time and place. Italy was politically fragmented and unstable; its petty states were constantly at war with one another, and it was often overrun by its powerful neighbors. On the other hand, Italy had some of the greatest trading centers of Europe. It was the birthplace of modern banking, and it was one of the two great industrial areas of Europe at the time (the Netherlands being the other). Italy also had the good fortune to be close to Byzantine and Islamic civilization. Italian engineering was also stimulated by the establishment of some of the world's first research centers, where under the sponsorship of men like Prince Francesco Sforza, engineer-architects could develop their talents while working on projects initiated by their patrons.[3]

Whatever the causes, by the end of the 15th century Italian engineers enjoyed an unparalleled reputation and were much in demand throughout Europe. For the most part they were engaged in military engineering, now made substantially more complex by the introduction of firearms, crude though they were. At the same time, Italian engineers became increasingly involved with civil works; they gave to Europe the mitred canal lock and the truss bridge, both designed to meet civilian rather than military requirements.

Their training was largely done through some form of apprenticeship, as it had been through the centuries. Still, we do find a few university graduates who combined engineering with mathematics or medicine, in the latter case maintaining a tradition that went all the way back to Imhotep. The training of the Italian Renaissance engineer did have one novel feature. Many Italian engineers went into the field after having apprenticed to an artist, sculptor, or goldsmith rather than a practicing engineer. Indeed, they often combined a career in engineering with one in the arts, Leonardo da Vinci being the prime example.

Accordingly, the Renaissance engineer frequently was not only the engineer-architect of earlier time, but an engineer-architect-artist, his work best exemplifying the relationship between technology and the arts that has always existed in some form.[4] The influence of these artist-engineers spread along with the rise of national monarchies throughout Europe during the 15th and 16th centuries as the rulers of newly consolidated states vied with their rivals to make their courts centers of culture. They also had the authority and resources to build powerful military forces and to promote extensive programs of public works. Thus Leonardo da Vinci spent part of his career in the court of the king of France, while some of his less famous compatriots enjoyed royal patronage in lands ranging from England to Russia.

Engineers of the Early Renaissance

The classical tradition that merged the role of the architect and the engineer is embodied in the career of Filippo Brunelleschi (1377-1446). He was a Florentine, the son of a successful lawyer who did much business with the military: handling their money, arranging for their pay, and even going on recruiting missions to northern Europe.[5] The elder Brunelleschi had expected his son to follow the same career, but when Filippo demonstrated artistic talent and asked to be apprenticed to a goldsmith, the father wisely consented. While working as a goldsmith, Brunelleschi took on an auxiliary trade: clockmaking. The practice of that craft was accompanied by an enthusiastic study of "motion, weights, and wheels, how they may be made to revolve and what sets them in motion."[6] Brunelleschi was thereby connected to one of the great engineering activities of the late Medieval world.

Despite his clockmaking skills, Brunelleschi is known principally as an architect, his principal achievement being the completion of the great cupola on the Florentine cathedral of Santa Maria del Fiore. The engineering required by this project was anything but trivial; the dome's span of 140 feet made it considerably larger than anything that had been attempted previously, and it was not to have any visible buttresses or other support. There is some dispute about how much Brunelleschi himself contributed to the architectural concept, but it is certain that in overseeing the actual execution he displayed marked engineering ingenuity in his technique for vaulting and in designing machinery for hoisting his materials into place.[7] Reinforcing Brunelleschi's claim to being an engineer is the fact that he seems to be the first person to receive a patent; the city of Florence awarded him one in 1421 for the design of a canal boat equipped with cranes.[8]

More important for our purposes, the record of Brunelleschi's career allows us some insight into how major construction projects were managed and financed, at least in the early years of the Renaissance. The building of the cathedral was overseen by a committee of representatives of the Florentine guilds, known as the *Opera del Duomo*. It provided the funds, approved plans and designs, hired and fired the architect-engineers, and bought materials, all this over a long period of time and with numerous changes of personnel.[9] Brunelleschi was probably brought into the project through his father's influence, since the senior Brunelleschi was an advisor to the *Opera*, possibly a member for a time. If so, it was case of nepotism that had happy consequences.

One of Brunelleschi's acquaintances was Mariano di Jacopo, known as Taccola.[10] He was born in Siena (the date is unknown, but he was baptised in 1382), the son of a wine dealer. He too was as much an artist as an engineer, and is credited with some of the sculptures found in Siena's cathedral. He also served as a notary and held a number of other public offices in his native city. He probably engaged in some engineering projects, but his most important legacy is not to be found in his actual creations. His most significant contribution to engineering consists of two notebooks, *De Ingeneis* and *De Machinis*. In these books he recorded technical details of a considerable range of structures and machines: bridges, buildings, mills, pumps, harbors, trebuchets, siphons, cofferdams, winches, hoists, surveying instruments, and water works. His books also included information culled from Byzantine writers on gunpowder and incendiaries. His notebooks show some knowledge of Heron of Alexandria, but most of the classic treatises remained undiscovered in his time, so there is no mention of Vitruvius or Euclid.

Despite its omissions, Taccola's notebook does mark the beginning of the modern technical treatise. It is based upon direct observation and the reports of practitioners, and is motivated by an evident desire to improve the speed, efficiency, and power of machinery.[11] Written before the invention of printing, Taccola's notebook never achieved wide circulation, but manuscript copies of it were well-known and found a place in the libraries of several prominent engineers. In this way it helped to pierce the veil of secrecy that had surrounded the work of late medieval architect-engineers who attempted to confine knowledge to fellow members of their guilds. As sources of information and inspiration, Taccola's notebooks marked the beginning of an era in which engineering knowledge was more readily available and eagerly consumed.

Taccola seems to have come from a family of modest means. By contrast, Leon Battista Alberti (1404-1472) came from a wealthy Florentine family (although he was born in Genoa, to where his family had been exiled). He studied at Padua and then at Bologna, where he started in law but then shifted to mathematics and science. Like Taccola, Alberti represented a significant new trend in engineering, for he too was the author of a book on engineering, *De Re Aedificatoria*. This was not the first book on structural engineering; a few treatises on the subject began to appear in late Medieval times. But they differed radically from Alberti's in the way they explored the subject. Medieval writing on engineering "...shows only what things can be done and how they should be done; it makes no attempt to explain to the reader why they have to be done in this particular way, let alone supply him with a system of general concepts on the basis of which he may cope with problems not yet foreseen by the writer."[12] In contrast, Alberti's book presented precepts based on general principles, coupled with a systematic approach to architecture and building. At the same time, Alberti's writings undoubtedly benefited from his work as an architect-engineer, for he was no armchair theoretician. He worked in many areas of architecture and engineering, and was conspicuously successful in the recovery of the remains of Roman ships from Lake Nemi.

Leonardo da Vinci

Brunelleschi, Taccola, and Alberti each came close to the ideal of the Renaissance Man; its culmination was Leonardo da Vinci (1452-1519). Enough has been written about him so that it is needless, and indeed impossible, to do more than summarize his career here. He was born in the town of Vinci, the illegitimate son of a Tuscan nobleman. At the age of twelve he was apprenticed to a Florentine named Verrochio, who was a painter, sculptor, and metal founder. He apparently picked up some mathematics during his apprenticeship, but most of his engineering was self-taught.

His emergence as an engineer can be dated from his famous letter of 1482 to Ludovico Sforza, the Duke of Milan. It was purely and simply and application for a job; any personnel manager would recognize it instantly as a resume. It is worth quoting in full, for what it tells about the state of the art in engineering in Leonardo's day:[13]

> Having, my Most Illustrious Lord, seen and now sufficiently considered the proofs of those who consider themselves masters and designers of instruments of war, and that the design and operation of said instruments is not different from those in common use, I will endeavor without injury to anyone to make myself understood by your Excellency, making known my own secrets, and offering thereafter at your pleasure, and at the proper time, to put into effect all those things which for brevity are in part noted below and many more, according to the exigencies of the different cases.
>
> I can construct bridges very light and strong and capable of easy transportation, and with them pursue or on occasion flee from the enemy, and still others safe and capable of resisting fire and attack, and easy and convenient to place and remove; and methods of burning and destroying those of the enemy.
>
> I know how, in a place under siege, to remove the water from the moats and make infinite bridges, trelliswork, ladders, and other instruments suitable to the said purposes.
>
> Also, if on account of the height of the ditches, or of the strength of the position and the situation, it is impossible in the siege to make use of bombardment I have means of destroying every fortress or other fortification if it is not built on stone.

I have also means of making cannon easy and convenient to carry, and with them of throwing out stones similar to a tempest; and with the smoke from them of causing great fear to the enemy, to his grave damage and confusion.

And if it should happen at sea, I have the means of constructing many instruments capable of offense and defense, and vessels which will offer resistance to the attack of the largest cannon, powder, and fumes.

Also, I have means by tunnels and secret and tortuous passages, made without any noise, of reaching a certain and designated point, even if it is necessary to pass under ditches or some river.

Also, I will make covered wagons, secure and indestructible, which entering with their artillery among the enemy will break up the largest body of armed men. And behind these can follow infantry unharmed and without any opposition.

Also, if the necessity occurs, I will make cannon, mortars, and fieldpieces of beautiful and useful shapes, different from those in common use.

Where cannon cannot be used, I will contrive mangonels, dart throwers, and machines for throwing fire, and other instruments of admirable efficiency and not in common use; and in short, according as the case may be, I will contrive various and infinite apparatus for offense and defense.

In times of peace I believe that I can give satisfaction equal to any other in architecture, in designing public and private edifices, and in conducting water from one place to another.

Also, I can undertake sculpture in marble, in bronze or in terracotta; similarly in painting, that which it is possible to do I can do as well as any other, whoever it may be.

Furthermore, it will be possible to start work on the bronze horse, which will be to the immortal glory and eternal honor of the happy memory of your father, My Lord, and of the illustrious House of Sforza.

And if to anyone the above-mentioned things seem impossible or impracticable, I offer myself in readiness to make a trial of them in your park or in such place as may please your Excellency; to whom as humbly as I possibly can, I commend myself.

Of the thirty-six abilities that Leonardo listed, thirty are concerned with technology and only six with the arts, lending credence to the claim of one commentator that Leonardo was "an engineer who occasionally painted a picture when he was broke."[14] The letter must have struck a responsive chord, for Leonardo got the job, as "ingenarius et pinctus." He stayed in Milan until 1499. While he was there he worked on the canals in the Po Valley that Milan had been developing since the twelfth century. Later, Leonardo served in a similar capacity for the city states of Venice and Florence, Cesare Borgia, and finally the king of France, where he was considered "premier pinctre et ingenieur et architecte du Roy."

The record is quite positive on Leonardo's status: a great artist beyond doubt, but at least equally eminent as an engineer. Leonardo's engineering interests ranged from fortifications to hydraulics, and his efforts at designing mechanized methods of textile production represent one of the first of the era's efforts to improve a manufacturing process.[15] Leonardo's interest in scientific discovery also displayed a distinct engineering mentality. As his narrative on hydraulics clearly states, "If you coordinate your notes on the science of the motion of water, remember to write beneath each proposition its applications, so that this knowledge does not remain unused."[16]

Leonardo's technical talents were formidable; two modern authorities who have studied his life and work have called him "the greatest engineer of all times."[17] There is no denying his imagination and virtuosity, but a conclusive evaluation of Leonardo as an engineer presents a number of problems. Many of the engineering creations that appear in his notebooks were actually the work of others.[18] Others were paper projects that were never translated into actual devices. Although Leonardo accomplished much as a military and civil engineer, many of his greatest visions were lost for at least three centuries, and some did not see the light of day until the 1960s. As a result, his influence on engineers and engineering was considerably less than it might have been.

Some of his admirers have made Leonardo the inventor of the airplane, the self-propelled armored vehicle, the helicopter, and other modern devices. This is surely stretching the meaning of "inventor." He had remarkable insights into how these devices might be made to function, but there is more to invention than drafting a design or even building a model to illustrate a conception. Leonardo faced the insurmountable handicap that many of his ideas were simply beyond the materials and skills of his day. He could conceive of a helicopter and even draw a seemingly plausible design, but it could not possibly have been built in the Europe of his day. And occasionally he committed a technical blunder; it has been pointed out that the gear train in Leonardo's design for a self-propelled vehicle was so arranged that the wheels would have rotated in opposing directions. Leonardo was also led astray by the idea of the ornithopter; flapping-wing devices have proven to be completely unsuited for artificial flight.

The above comments may seem a bit mean-spirited. But they are only intended to provide some perspective. Leonardo's genius is unquestioned, but in terms of what he actually accomplished as an engineer he has to be viewed in the context of his times. He was one of a number of brilliant early Renaissance engineers, just as he was one of a number of brilliant artists who emerged in Italy at that time. His genius was distinctive, but he lacked many of the things taken for granted by today's engineers. The Renaissance provided a supportive setting for the development and exercise of genius, but centuries were to pass before brilliant insights could be transformed into actual accomplishments.

The Efflorescense of Italian Engineering

Leonardo's genius did not emerge in a cultural vacuum. The Italian engineers just described had contemporaries and near-contemporaries with equally noteworthy accomplishments. And after Leonardo there came others who made substantial contributions to engineering in their native Italy and throughout Europe.[19]

Probably more prominent and influential as an engineer than Leonardo was Francesco di Giorgio (1439-1502). He was a native of Siena and is identified as coming from a family of modest means. He too had an artistic training and practiced sculpture throughout his life, a vocation that led him to become a founder of bronze. Emerging military technologies led to his making bronze cannon barrels instead of bronze statuary. Like many other military engineers, his talents spilled over into the civilian realm; by the time he was thirty he had won acclaim by maintaining the water supply of Siena. He also played an important role in the transmission of engineering knowledge; between 1470 and 1480 he completed a treatise on architecture and engineering which came close to being a handbook for the profession. Di Giorgio had some ingenious ideas about machines, especially in regard to gearing, but as with Leonardo, he was limited by the inadequacy of the technology of his time. His designs for suction and pressure pumps and for a hydraulically powered saw were feasible in principle, but their realization was stymied by the limitations of metallurgy and metal working and the availability of proper lubricants.[20]

The prevailing system of apprenticeship lent itself to skills being passed from father to son, and engineering was no exception. The Italian Renaissance produced several families of distinguished engineers. Ridolfo Fioravanti (?-1479) was the son of a man who had worked on the canals around Milan but about whom little is known otherwise. Ridolfo also worked on these canals, and later he engaged in some difficult architectural work. He subsequently went to Hungary and then to Russia, where he served as an architect, gun maker, bell founder, and head of the mint. Other Italian architect-engineers followed him to Russia. One cannot help but wonder if their first exposure to the Russian winter gave rise to the thought that the transfer of technical skills through migration was coming at too high a price.

The Milanese canals that the Fiorvantis worked on deserve some discussion. The system was begun late in the 12th century, chiefly to bring water from the Ticino River for irrigation and to supply water for the moat encircling the city. In the mid-fifteenth century the Sforza dynasty undertook to expand the canals so as to carry commercial traffic over a wide area. The resulting effort employed some outstanding engineers, most notably Leonardo, and like the Erie canal some centuries later, it served as a school for training engineers. In 1452 Bertola di Novate was put in charge of the work. One modern writer hails him as "the outstanding hydraulic engineer of the second half of the century" and credits him with construction of the first canal locks[21] but regrets that very little is known of his antecedents or his career. He was a citizen of

Milan, and, significantly, he was always listed as an engineer — not as an architect or engineer-architect.

The canal's ultimate purpose was to connect Milan with Lakes Como and Maggiore on the north and the Po River to the south. The terrain presented many serious difficulties, and required the deployment of canal locks to accommodate changing water levels. This proved to be beyond the technical skills available to di Novate, so that he was able to complete only part of what he had planned. Then the work was further delayed by warfare between France and Spain for control over Northern Italy.

Subsequent efforts to complete the system were made in the 16th century by a number of engineers. The most significant was another Milanese, Giuseppe Meda. Little is known of his background except that his father was an architect and that he himself was a painter before turning to engineering. During the last quarter of the sixteenth century Meda developed ambitious plans for completing the canal system, including the design of a single massive lock to take the place of a multitude of smaller ones. But Meda fell prey to problems that occasionally bedevil modern engineers: labor disputes, cost overrruns, and lawsuits. As a result, he landed in prison. In 1599 the report of another engineeer subsequently vindicated him and his work, but his health was broken and he died shortly after his release. The impetus to finish the canal system was lost, not to be regained until the 19th century.[22]

Another important family of Renaissance engineers were the San Gallos. The father, Francesco Giamberti (dates unknown) was a carpenter and cabinet maker who eventually became an architect. The name San Gallo was taken by his sons Giuliano (dates unknown) and Antonio (1455-1535). The two brothers achieved a reputation as military engineers, but their most noted achievement was to move Michelangelo's statue of David from the workshop to its present site in Florence.

Another effort at moving a work of art was one of the most heralded feats of Renaissance engineering. It was undertaken by Domenico Fontana (1543-1607), who already had a considerable reputation as an architect when he was given the task of moving an Egyptian obelisk to St. Peter's square from the place where the Romans had placed it in the first century A.D. Since the obelisk weighed about 340 tons and Fontana had no power machinery at his disposal, accomplishing this task was a spectacular feat for its time (to be sure, the ability of the Romans to move the obelisk from Egypt to Rome hundreds of years earlier was even more impressive).

Fontana was undoubtedly a highly competent engineer, but it certainly did not hurt that he enjoyed the patronage of Pope Sixtus V, for whom he had worked when the Pope was still a cardinal. Papal sponsorship allowed Fontana to free himself of any worries about the non-engineering aspects of the job. He had power over just about everything connected with the project; he could requisition lumber and ropes, as well as pasturage for his draft animals, and could even exercise eminent domain in order to tear down buildings that stood in the way of his equipment as it was moved through the city. Compensation had to be paid for all these things, although it seems unlikely that the other parties got the better of the deal.[23]

Feats such as those of Fontana captured the imagination of the public, but a more important legacy of the Rennaisance era was the successful attempt to put engineering on a firmer scientific and mathematical basis. One early example was the work of Nicolo Tartaglia (1500-1557). He came from a family of modest means, and only with considerable difficulty was he was able to attend a school that taught reading. But lack of money forced him to leave school before he had learned the full alphabet, and he had to learn all the letters past "K" on his own. He also devoted himself to the self-study of mathematics, using what he had learned to produce the first vernacular translation of Euclid. On a more practical level, he applied his mathematical knowledge to the study of ballistics, a field that was taking on special importance as artillery became increasingly prominent on the battlefields of Europe. Not satisfied with the prevailing rule-of-thumb procedures, Tartaglia conducted research on the flight of cannon balls by combining experimentally derived data with theoretical reasoning.[24] As a result of his efforts, artillerymen learned that a cannonball travels in a curved path, and that a 45-degree inclination of the barrel gives the greatest distance to the projectile.[25]

A generation after Tartaglia demonstrated the practical value of mathematics, another Italian engineer, Agostino Ramelli (?1531-?), expressed his great enthusiasm for that subject: "...there cannot be found nor can there be, among the liberal arts, any science more noble or illustrious than that of mathematics, in which there truly appears to be present some indwelling power of unfathomable divinity."[26] But unlike Tartaglia, Ramelli did little to advance the fusion of mathematics and engineering. Rather, his greatest accomplishment was the publication in 1588 of a book illustrating a great variety of mechanical devices. His book can be seen as a continuation of the work begun by Taccola, and it provided a foundation for several subsequent works.[27] Ramelli's book contains nearly two hundred engravings depicting, in beautiful detail, machines for raising water, milling grain, and moving and lifting heavy objects. It also shows the importance of military matters through pictures of screwjacks for tearing down doors and gates, temporary bridges for crossing moats, and catapults for throwing heavy weights.

Ramelli's praise for mathematics notwithstanding, his book provides no calculations for determining such key variables as the size of component parts, gear ratios, or the strength of materials employed. But Ramelli's book must have helped to generate an interest in mathematics. It presents such a vast array of machines that could be used for a single purpose, such as lifting water, that a practically minded reader could not help but wonder how to choose between them. It seems reasonable to suppose that this problem helped to stimulate systematic efforts to evaluate the efficiency of different machines, thereby paving the way for the more extensive use of measurement and calculation.

The engineers of Renaissance Italy seem to be a diverse lot, although information about their social origins remains fragmentary. Of the 59 Italian engineers mentioned in Bertrand Gille's *Engineers of the Renaissance*, no information is available on family background for forty-five of them. Of the others, seven came from professional or mercantile families, four were craftsmen's sons, one was the son of a landowner, one the son of a farmer, and one, Nicola Tartaglia, grew up in poverty. Information about their education is equally lacking. For thirty-three there is no indication of how they were educated or trained. Eighteen had some form of apprenticeship, three were university products, and six were self-taught. The information is sketchy, but it appears clear enough that the engineer-architect-artist of the Italian Renaissance was a respected figure in his culture and that he worked in a field where entrance was determined at least as much by talent as it was by prior social and economic status.

The Engineers of Northern Europe

The development of an engineering profession in northern and western Europe was powerfully influenced by the Italian Renaissance, but it also had its indigenous origins. In the Netherlands the constant struggle against the encroachment of the sea provided training for successive generations of engineers. More important, the Netherlands was developing into a center for art and science comparable to Italy. By the early seventeenth century, it was an independent republic, noted for its religious toleration, along with being the center of the Western world's commerce and banking, and the home of towering figures in the arts, sciences, and law. Yet while the names of Erasmus, Rembrandt, and Grotius have been immortalized, the engineers who provided the material underpinnings of the golden age of Dutch civilization have remained largely anonymous.

One of the few who gained a measure of fame was Simon Stevin (1548-1620). Stevin had a most varied career. He began as a cashier in a mercantile house in Antwerp, travelled widely, and studied literature at the University of Leyden. Later he became inspector of dykes, taught mathematics at the Leyden School of Engineering (which shows how the field was changing), and finished his career as Internant of the Dutch Army.[28] As an engineer he built water mills and worked on various drainage problems— major concerns in a country whose land often lies below sea level.

Stevin's accomplishments as a scientist and mathematician were at least the equal of his engineering successes. It was Stevin who actually performed the experiment that legend attributes to Galileo when he dropped iron balls from a tower to prove that the weight of a body does not affect the velocity at which it falls. He also produced the first tables of compound interest extensive enough to be used by bankers, and was the first European to use decimal fractions. Equally important was Stevin's enlistment of mathematics into engineering practice. He was, in fact, one of the first engineers to consciously apply advanced mathematics to engineering problems. This represented a new departure for mathematics also. Because he was a practicing engineer, his mathematics did not represent a quest for abstract "pure" knowledge, but was wielded as a practical tool.[29]

For the period immediately following the Renaissance, probably the best known Dutch engineer was Cornelius Vermuyden (1595?-1683?) He was born on the island of Tholen in Zealand and at an early age became engaged in draining and reclaiming land. His father was probably an engineer also. Vermuyden's reputation was such that when the Thames broke its banks below London in 1621, he was summoned to England to take charge of repairs. He spent the rest of his life in England as both an engineer and entrepreneur engaged in drainage projects in the Fenlands of eastern England, receiving his remuneration primarily in grants of portions of the lands to be reclaimed.

These enterprises embroiled Vermuyden in endless controversy. He was plagued with lawsuits from disgruntled landowners, and there were also strong concerns about what his projects were doing to the local environment. As a result, his workers were occasionally attacked by inhabitants of the region who lived by fishing and hunting the wildlife that the fens sheltered in great numbers. Making matters worse, the Dutch workmen that he brought over neither liked nor were liked by the local populace. Even his engineering methods were criticized. He stayed sufficiently in favor to be knighted in 1628, but eventually the constant litigation and the limited success of his efforts compelled him to sell his landholdings, and he died in poverty. The fault was not necessarily his own. The actual engineering was difficult enough, but far more problematic were the circumstances he worked in. The Fenland drainage projects of this era were a tangle of conflicts over local rights, rivalries among landowners, and the role of the Crown, which was itself an important Fenland landholder.[30] These circumstances could defeat the most technically proficient engineer.

Vermuyden's English contemporary, Sir Hugh Myddleton (1560?-1631), fared rather better. He was born in Denbigh, Wales, the sixth son of Richard Myddleton (or Middleton), the governor of Denbigh castle and a member of Parliament, which seems to put him among the lesser gentry. Hugh Myddleton was apprenticed to a goldsmith, and remained in banking, an allied field, throughout his life.[31] He was also a successful merchant, ran a clothmaking business, and was Member of Parliament for Denbigh for many years. His entry into engineering came in 1609, when he took charge of a proposal for bringing fresh supplies of water to London. The project was completed in 1613. It consisted of a canal 38 miles long, ten feet wide, and four feet deep that brought water from Hertfordshire to a reservoir in Islington. The canal, known as the New River, is still part of London's water supply system; it has been improved and expanded, but is still based on Myddleton's work.

The New River operation left Myddleton heavily in debt, but he was able to recoup by taking over some lead and silver mines in Wales that had been abandoned because of flooding. He succeeded in draining them and making them profitable, and he also had some success in reclaiming land on the Isle of Wight. In recognition of his engineering feats he was granted a baronetcy in 1622. This award, like Vermuyden's knighthood, was an unusual recognition for an engineer; the next engineer to be knighted was John Rennie, Jr., more than two hundred years later. The engineer may have been regarded a useful person to have around, but in the England of the Tudors and Stuarts he was likely to be sent to the tradesman's entrance.

There is even a suggestion that the hostility of literary intellectuals to technology, as described by C.P. Snow in *The Two Cultures*, antedates the Industrial Revolution. Andrew Marvel, a poet who served as secretary to both John Milton and Oliver Cromwell, and in some diplomatic posts after the Restoration, had this to say about the government of the Dutch Republic:[32]

> So rules among the drowned he that drains;
> Not who first sees the rising sun, commands,
> But who could first drain the rising lands;
> Who best could know how to pump and earth so leak,
> Him they their Lord and Country's Father speak;
> To make a bank, was a great plot of state,
> Invent a shov'l, and be a Magistrate.

In expressing his amused contempt for a people who could put engineers in its highest offices, Marvel is inadvertantly telling us something important about Dutch society: engineering was a prestigious occupation in the Netherlands.

The Mining Engineer in Germany

German engineering was concentrated on mining, an industry that had grown up around the rich deposits of gold, silver, lead, copper, iron, tin, and nickel of southern Germany. The German mining engineers have fared no better in history's records than any others, except in one respect. The results of their efforts were recorded in one of the great classics of technological writing, *De Re Metallica* by Georg Bauer (1494-1555), or the Latinized name by which he is usually known, Giorgius Agricola. Agricola was not an engineer himself, but he was a well-educated man with a wide range of interests. He was born in Saxony, one of the principal mining areas in Germany. He attended the University of Leipzig and various Italian universities, and he became a friend of Erasmus. He studied medicine in addition to literature and philosophy, and became a physician in Joachimsthal, then a booming Saxon mining town.

Agricola wrote in considerable detail about the mining operations around him, including in his treatise mining law, ore formations, pumps, hoisting machines, the use of water power, the manufacture of various chemicals, and metallurgy. Some of his descriptions seem to have been based on an earlier publication, *De la Pirotechnia* (1540) by Bannoccio Biringuccio (1480?-1539) of Siena, who was both a mathematician and metallurgist. Agricola's work shows a level of mining technology that was not substantially changed until the introduction of power machinery and dynamite in the 19th century.[33] The first good English translation of *De Re Metallica* was published in 1912 by a classical scholar, Lou Henry Hoover, and her husband, Herbert Hoover, the same Hoover who went on to an ill-starred presidency after a successful stint as a mining engineer.

Like so many other writers on engineering, Agricola was interested in devices and techniques, not people, so he does not identify any specific engineers. However, there are occasional passages that described the kind of person who was in charge of various operations and who presumably would now be considered engineers or metallurgists. In his description of how mining operations were conducted in Saxony and Thuringia, Agricola identifies an official known as the mining prefect, who was the direct representative of the king or prince, and had full authority to regulate mining.[34] Under him was the Bergmeister, who was granted the right to mine in his area, gave instructions to mine owners, kept records, supervised operations, and settled disputes.[35] Disputes that he could not settle were referred to the mining prefect. Such officials obviously had to have a thorough understanding of mining technology, and today would be classified as mining engineers.

Engineers in France

The importation of Italian engineers into France during the height of the Italian Renaissance has been noted earlier in this chapter. Later, engineers from the Netherlands played the same kind of role in France that they did in England, working on drainage and hydraulic projects. One of them Humphrey Bradeley (or Bradley) achieved distinction in both England

and France. He was born in Bergen-op-Zoom, probably the son of an English merchant who was stationed there as an official of the Merchant Adventurers.[36] The date of his birth is unknown, but he appeared in England in 1584 to work on the reconstruction of Dover harbor. Four years later he was asked to make a study of the Great Level, the largest segment of the Fens. The report that he and some associates submitted was the first and for a long time the only systematic discussion of how to convert the Fens into productive farmland. It appears to have influenced Vermuyden quite strongly.

However, Bradley's proposals received no support at the time, and in 1593 he returned to the Netherlands and three years later went to France in response to a request from Henri IV for "individuals qualified in the art of diking."[37] He became master of the dikes under Henri. About 1600 he also submitted a proposal to connect the Rhine and the Seine with a canal, and formed a syndicate to carry out the plan. This project failed to get the needed financial support, so the scheme was not carried out.[38] A canal following Bradeley's approximate route was finally built in 1775.

Another Netherlander, Jean Lintlaer, was brought to Paris in 1602 to construct a water supply for the royal palaces in the city, the Louvre and the Tuileries. Although most of the details have been lost, we do know that his design used a water wheel-driven pump under the arches of the Pont Neuf.[39] An objection from Parisian city officials that this work might interfere with navigation on the Seine was brusquely rejected by the king, who argued that the placement of a water wheel was none of their business, for his money had paid for the bridge in the first place.[40] Having the king for a patron could certainly make an engineer's job easier when it came to the nontechnical aspects of the operation.

Even as Brindley and Lintlaer were going about their work, France's reliance on imported engineers was rapidly disappearing and the groundwork was being laid for the great flourishing of French engineering that occurred in the eighteenth century. Much of the stimulation for the early growth of French engineering came from the great minister of Henri IV, Maximilien de Bethune, duc de Sully (1560-1641), who had himself practiced engineering before devoting himself to public office. With France's long and exhausting wars of religion finally at and end, Henri and Sully threw themselves energetically into rebuilding their country. One result was a surge of bridge and canal building in which the bulk of the engineering skill was French. After the assassination of Henri in 1610 there was another period of weak government and internal unrest, along with involvement in the Thirty Years' War (1618-1648). These delayed some of the projects, but the return of stability brought a revival in the second half of the century.

The bridge builders were architects (sometimes calling themselves architect-engineers) or master masons and carpenters. It is interesting, but hardly surprising, that there was still no essential distinction between the architect and the engineer; the practitioners were basically concerned with structures. The names of some of them are known, but not much else. More information is available about the principal canal builders of the period. The earliest was Adam de Craponne (1526-1575) who built a large irrigation canal in Provence, near Arles. The family came originally from Italy, but had settled in France for at least two generations before Adam was born. He became a military engineer of some distinction before he was authorized to undertake the canal. His work was another example of an engineer taking on the role of entrepreneur, for the work was to be done at his own expense.[41] And like all too many engineer-entrepreneurs of his era, he lost heavily on the project.

Craponne also studied the proposed Charollais Canal that was intended to connect the Loire and the Rhone and thereby provide a direct route between the Atlantic and the Mediterranean. He later decided that the Languedoc route subsequently followed by another engineer, Pierre-Paul Riquet, was better suited for the purpose. Craponne died when he was sent to investigate the failure of the foundations of the fortifications at Nantes. His demise seems to be the result of his efforts to fix the blame for shoddy construction. Allegedly, the contractors "sought to purchase him by presents and flattery, but as they could not succeed by such means, removed him by poison."[42]

The Briare canal, intended to link the fertile Seine and Loire Valleys, was directly instigated by Sully. It was an attractive but difficult proposal, for it had to cross a watershed that rose 400 feet above the Seine. The work was entrusted in 1604 to Hugues Cosnier, described by the distinguished British engineer-historian L.T.C. Rolt as an engineer of genius and remarkable tenacity of purpose.[43] Difficulties developed over technical points concerning the route to be followed, and in 1606 Sully took over and relocated the canal, at the same time providing for larger locks. The task was returned to Cosnier a year later but disputes continued. When the assassination of the King in 1610 ended Sully's influence, work was suspended until Cardinal Richelieu revived the project in 1639, and the canal was completed four years later.

Finally came the Languedoc canal, or the Canal du Midi, unquestionably the most striking feat of 17th century French engineering. Rolt goes further, and describes it as "undoubtedly the greatest work of engineering in the world at that time. Nothing on so grand a scale had been built by man since the fall of the Roman empire."[44] This judgment betrays the Europeocentric tendency of many historians of technology; the Canal du Midi was an impressive engineering accomplishment, but it pales beside China's Grand Canal.[45] Completed more than 350 years before the Canal du Midi, the Grand Canal stretches from Hangzhou in the South to Tianjin in the North, a distance of nearly 1100 miles, making it one of history's greatest works of civil engineering.

The Grand Canal was constructed primarily to transport tax payments in the form of grain from the South to the Imperial capital. The Canal du Midi was built to facilitate commerce. The different purposes for which the two canals were built serve to exemplify the fundamental difference between bureaucratically controlled China and the emerging bourgeois society of Europe. The idea of connecting the Mediterranean with the Atlantic with a waterway so that traffic did not have to make the long and sometimes perilous journey around the Iberian peninsula had a strong appeal in France at a time when inland transportation was severely limited by the roads of the time. The most direct route was approximately 150 miles in length and encompassed a vertical distance of 621 feet from the summit to sea level.

What today would be called feasibility studies were initiated early in the 16th century. The reports were favorable, but it was quite obvious that the scheme would tax French financial resources and engineering skills to the limit. The outbreak of a series of civil wars put the project on the shelf for the rest of the century until it was revived during the second half of the seventeenth century by Jean-Baptise Colbert, the finance officer of Louis XIV. In 1662 Pierre-Paul Riquet de Bonrepos (1604-1680) offered to undertake the canal, and with Colbert's support gained the necessary approval of the local authorities as well as some financial assistance.

Riquet was apparently descended from an Italian family named Arrighetti who had moved to Southern France in the 13th century. His grandfather was a master tailor who married a wealthy woman, and his father was a successful lawyer who became a magistrate in the city of Beziers. Pierre-Paul himself was educated at the Jesuit college in the city of Beziers, where he did his best work in mathematics.[46] He too married into a wealthy family, and was able to purchase the estate of Bonrepos. He also became the collector of the salt tax for the province of Languedoc, a lucrative if highly unpopular position. When he undertook the canal it was in the capacity of entrepreneur. Like the builders of the Erie canal later on, he was not an engineer when the work started but definitely was well before it finished.

Riquet received invaluable professional assistance from a trained engineer, Francois Andreossy, a Parisian who had moved to Narbonne. Andreossy had been educated at Paris as a civil engineer, with a special interest in hydraulics, and when Riquet began to plan the Canal du Midi he had just returned from a trip to study the canals of Northern Italy.

The Canal du Midi was begun in 1667 and opened in 1681, a year after Riquet's death. It was 148 miles long, had a hundred locks and seven aqueducts, as well as a 500-foot tunnel, which was the first in which gunpowder was used for blasting.[47] Riquet was left impoverished—a sadly recurring theme with the engineer-entrepreneurs of this era—but some forty years later his heirs began to make money from the canal.

The Legacy of the Renaissance Engineers

By this time the Renaissance was well in the past. The era had been one of great dynamism and many accomplishments, but its technological achievements were more evolutionary than revolutionary. Indeed, it can be argued that Renaissance technology was not a distinctive creation, but essentially an elaboration and refinement of what had been inherited from the Middle Ages.[48]

All the same, the practice of engineering was undergoing significant changes. While a Medieval master mason may have found illiteracy to be no obstacle to success, the engineers of the Renaissance and after were almost universally literate. This of course was a reflection of the generally higher levels of literacy for the entire populace, but for the engineer it had a particular importance. From the beginning of the fifteenth century onward, a veritable flood of technical literature swept through Europe.[49] Ranging from detailed how-to-do-it manuals to rambling reflections, the technical literature of the time obviated a considerable amount of trial and error, and made some on-the-job training superfluous. Engineering was still far from being a common university subject, but it was slowly bridging the gap that separated the world of the craftsman from that of the scholar.

Reinforcing this trend was the slow but steady injection of science into engineering. Perhaps "injection" is the wrong word, for a great deal of science was directly or indirectly the product of engineering practice. In any event, the work of the engineer was beginning to be guided by fundamentally different ways of thinking and working. In Gille's words, Renaissance "engineers... offered their period new ways of thought and untried methods. Reasoning took the place of empiricism, experiment replaced the rule of thumb, [and] scientific calculation ousted elementary relationships."[50]

In addition to bringing changes to engineering training and practice, the Renaissance era was significant in that it marked the demarcation of engineering as a distinct occupation. Unlike the Middle Ages, the designers and builders of structures and mechanisms were recognized, and recognized themselves, as engineers rather than master craftsmen of various kinds. Most of them, certainly the most prominent ones, were also architects, painters, or sculptors, and sometimes all three, but they saw their work as engineers as a separate and at least equally prestigious role. Most of their engineering continued to be done for military purposes, but with the passage of time came a noticeable increase in civil works, such as the French canals and the drainage projects of England and the Netherlands.

Although we have focussed on some of the most prominent engineers, the typical Renaissance engineers were not the Leonardos and Fontanas; rather, they were lesser figures who did not build great churches or produce great art, but concentrated on down-to-earth engineering. A complete picture of the Renaissance engineer would include a large number of rather obscure individuals, but individuals who were nonetheless clearly definable as engineers. They may not have reached the intellectual and aesthetic heights of their more exalted contemporaries, but collectively they provided the period with the technological underpinnings of sustained economic growth. Great and small, the engineers of the Renaissance helped to lay the groundwork for the tech-nological revolution that was to follow.

Notes

1 L. Sprague de Camp, *The Ancient Engineers* (Garden City, New York: Doubleday, 1963), p. 357.

2 A.G. Keller, "Mathematics, Mechanics, and Experimental Machines in Northern Italy in the Sixteenth Century," in Maurice Crosland (ed.), *The Emergence of Science in Western Europe* (New York: Science History Publications, 1976), p.16.

3 Bertrand Gille, *Engineers of the Renaissance* (Cambridge, Massachusetts, MIT Press, 1966), pp. 42-45.

4 Cyril Stanley Smith, "Art, Technology, and Science: Notes on Their Historical Interaction," *Technology and Culture*, 11, 4 (October 1970)

5 Antonio Manetti, *The Life of Brunelleschi*, trans. Catherine Engass (University Park: University of Pennsylvania Press, 1970)

6 The quotation is by Vasari, and appears in Paulo Rossi, *Philosophy, Technology, and the Arts in the Early Modern Era* trans. Salvator Attenasio (New York: Harper and Row, 1970) p. 18.

7 Frank D. Prager and Gustina Scaglia, *Brunelleschi: Studies of His Technology and Inventions* (Cambridge, Massachusetts: MIT Press, 1970), p. 6.

8 de Camp. op. cit., p. 359.

9 Prager and Scaglia, op. cit., p. 2.

10 Frank D. Prager and Gustina Scaglia, *Mariano Taccola and His Book, De Ingeneis* (Cambridge: Massachusetts: MIT Press, 1972)

11 Ibid., p. 153.

12 Erwin Panofsky, *The Life and Art of A. Durer* (Princeton: Princeton University Press, 1955), p. 242.

13 Ladislao Reti and Bern Dibner, *Leonardo da Vinci, Technologist* (Norwalk, Connecticut: Brundy Library, 1969), pp. 45-46.

14 Lynn White, Jr., *Medieval Religion and Technology: Collected Papers* (Berkeley: University of California Press, 1971), p. 38.

15 Gille, op. cit., p. 166.

16 Ibid., p. 189. Gille calls this orientation "unscientific," which hardly seems accurate or fair.

17 Reti and Dibner, op. cit., p. 96.

18 Jean Gimpel, *The Medieval Machine: The Industrial Revolution of the Middle Ages* (Harmondsworth: Penguin, 1977), p. 142.

19 Unless otherwise noted, all biographies in this section are based on Gille, op. cit.

20 Gille, op. cit., p. 111.

21 William B. Parsons, *Engineers and Engineering in the Renaissance* (Cambridge, Massachusetts: MIT Press, 1939), p. 370. Di Novate was undoubtedly a good engineer, but to credit him with the invention of the canal lock is to ignore prior work by Chinese engineers that antedated his invention by centuries.

22 Ibid., pp. 405-416.

23 Ibid., pp. 155-67.

24 A. Koyre, "The Exact Sciences," in Rene Taton (ed.), *The Beginnings of Modern Science* (New York: Basic Books, 1964), p. 90.

25 Friedrich Klemm, *A History of Western Technology* (Cambridge, Massachusetts: MIT Press, 1964), p. 131.

26 Agostino Ramelli, *The Various and Ingenius Machines of Agostino Ramelli* trans. Martha Teach Gnudi, with technical annotations by Eugene S. Ferguson (New York: Dover, 1987), p. 54.

27 An immediate predecessor of Ramelli's book was Jacques Besson's *Theatre des Instrumens Mathematiques et Mechaniques* (Lyon, 1578). Two of the most important subsequent works were Vittorio Zonca's *Novo Teatro di Machine et Edificii* (Padua, 1607) and Jacob Leupold's *Theatrum Machinarum* (Leipzig, 1724-39). The Islamic world also produced noteworthy books of this sort, such as Banu-Musa, *The Book of Ingenius Devices*, trans. Donald R. Hill (Dordrecht: Reidel, 1979) and al-Jazari, *The Book of Knowledge of Ingenius Mechanical Devices*, trans. Donald R. Hill (Dordrecht: Reidel, 1974).

28 Koyre, op. cit., pp. 46-47.

29 Gille, op. cit., pp. 222-23.

30 On the political and legal complexities of Fen drainage, see Margaret Albright, "The Entrepreneurs of Fen Draining in England under James I and Charles II: Illustrations of the Uses of Influence," *Explorations in Entrepreneurial History*, 8, 2 (1955)

31 *Dictionary of National Biography*, vol. 10 (London: Oxford University Press, 1973), pp. 1333-35.

32 Charles Wilson, *The Dutch Republic* (New York: McGraw-Hill, 1968), p. 43.

33 James K. Finch, *The Story of Engineering* (Garden City, New York: Doubleday, 1960), pp. 130-33.

34 Agricola, *De Re Metallica*, trans. Herbert Clark Hoover and Lou Henry Hoover (London: The Mining Magazine, 1912; originally published in 1556), pp. 94-95.

35 Ibid., pp. 84-90, 95-97.

36 L.E. Harris, "Land Drainage and Reclamation," in Charles Singer (ed.), *A History of Technology*, vol. 3 (New York: Oxford University Press, 1957), pp. 316-17.

37 Ibid., p. 318.

38 Parsons, op. cit., pp. 443-46.

39 Ibid., p. 249.

40 Finch, op. cit., p. 139.

41 Parsons, op. cit., p. 454.

42 Ibid., p. 455.

43 L.T.C. Rolt, *From Sea to Sea: The Canal du Midi* (Athens, Ohio: Ohio University Press, 1973), p. 20.

44 Ibid., p. 2.

45 Joseph Needham, *Science and Civilization in China*, vol. 4, part 3: *Civil Engineering and Nautics* (Cambridge: At the University Press, 1971), pp. 306-20.

46 Rolt, op. cit., pp. 28-31.

47 Richard Shelton Kirby, et al., *Engineering in History* (New York: McGraw-Hill, 1956), p. 141.

48 Herman Kellenbenz, "Technology in the Age of the Scientific Revolution 1500-1700," in Carlo M. Cipolla (ed.), *The Fontana Economic History of Europe: The Sixteenth and Seventeenth Centuries* (Glasgow: Collins/Fontana, 1974), pp. 264-65.

49 Rossi, op. cit., p. 15.

50 Gille, op. cit., p. 215.

Chapter 5

The Emergence of the Civil Engineer

It is a sad commentary on humanity's use of technology that so much of it has been applied to warlike ends. As a result, much of the stimulus for the development of engineering has come from the military sector. Individuals who were described as engineers were usually attached to armies, while the "engines" that they created were catapults and other martial devices. This situation began to change during the 18th century with the emergence of a new breed of engineer, men who practiced "civil" (i.e., non-military) engineering.[1] This in itself was an indication of the expanding role of the engineer in Western civilization. It also offered the happy prospect that to a greater degree engineering talent could be directed at something better than killing and destroying.

This chapter will concentrate on the rise of civil engineering in France and Britain. This emphasis should not be taken as an indication that no other countries produced civil engineers. Even a belief in the technological primacy of Europe as a whole is manifestly false, for as we have seen, no pre-industrial nation surpassed China in the sophistication and scope of civil engineering projects. We have chosen to focus on Britain and France simply because they were the leading forces in the development of engineering during the eighteenth and early nineteenth centuries. Equally important, France and Britain exhibited markedly different political and social climates, and these strongly affected the nature of engineers and engineering in the two countries. Whereas the growth of civil engineering in Great Britain was individualistic and unregulated, in France it was fostered and organized by the central government. An examination of British and French engineering will therefore help us to better understand how engineering is affected by its social and political context. Not all of the relevant issues will be explored in this chapter, but it will help to introduce some of the themes that will be considered in greater depth in subsequent chapters.

The Organization of Engineering in France

The organization of French civil engineering began, somewhat paradoxically, with one of the greatest military engineers of all time, Sebastian Le Prestre de Vauban (1633-1707). He was born in Burgundy into a family of the lesser nobility and became a military officer at the age of 17. A few years later he was admitted to a body of engineer-officers that was formally constituted, apparently at Vauban's suggestion, as the *Corps des Ingenieurs du Genie Militaire* in 1675.[2] *Genie* officers, including Vauban himself, were often assigned to civil works, much as with today's U.S. Army Corps of Engineers.

The influence of the *Corps du Genie* was not limited to the projects they built. At a time when formal instruction in engineering was virtually unknown, the Corps established educational institutions in various parts of France to train its officers. These early engineering schools had a substantial scientific component in their curriculum, and some of their students and faculty had distinguished careers in science and mathematics. The school at Mezieres was notable for having the eminent mathematician Gaspard Monge (1746-1818) on its faculty; he was the creator of descriptive geometry, one of the essential tools of modern engineering. The school also produced two very distinguished graduates. One was Charles Agustin Coulomb (1736-1806), who combined military and civil engineeering with pioneering research in electricity and magnetism. The other was Lazare Carnot (1753-1823), the "Organizer of Victory" in the wars of the French Revolution and also a mathematician of note. Perhaps even more important, he was the father of Sadi Carnot, one of the most important figures in the development of the science of thermodynamics. Both Coulomb and Carnot came from families of lawyers and public officials; their enrollment at Mezieres reflects the fact that under the Old Regime the *Corps du Genie* was the only branch of the French army where a commoner could get a commission.

In 1716 a non-military organization, the *Corps des Ponts et Chaussees*, was added to France's engineering establishment. The establishment of it and the *Corps du Genie* reflected France's need as a great land power to have good internal communications for both military and commercial reasons. Its graduates met this need by giving France the only good transportation network in 18th century Europe, a system that encompassed about 30,000 miles of roads and highways. The building of the road system entailed a substantial political cost, however. Its construction made heavy use of the *corvee*, the forced requisitioning of local labor. Universally despised by the peasantry, it was one of the grievances that eventually led to the French Revolution.

Two members of the *Corps des Ponts et Chaussees* deserve special credit for France's road-building achievements. Jean-Rodolphe Perronet (1708-1794) was the son of a Swiss officer in the French army. He intended to join the *Corps du Genie*, but the death of his father obligated him to enter an architect's office instead.[3] In 1735 he became an engineer in the *Corps des Ponts et Chaussees*, and eventually rose to be its head. His greatest achievements were in bridge design, especially the stone arch bridge, of which the Pont de la Concorde in Paris is a noteworthy example.

Pierre-Marie-Jerome Tresauguet (1716-1796) was a Parisian, the scion of a family of engineers. His road building methods consisted of laying successive courses of stones, beginning with flat stones laid on edge and hammered into the ground, and ending with a surface of small stones. These courses allowed adequate drainage without requiring a dangerously high camber. Tresauguet was thus clearly the initiator of the road-building techniques that in the English-speaking world have been attributed to Thomas Telford and John McAdam.[4]

With such a solid educational foundation, France was able to produce a succession of eminent engineers such as Perronet and Carnot. The curriculum of French engineering schools reinforced the emerging trend of giving engineering a stronger mathematical and scientific foundation. The basic mathematics text for engineers was the work of Bernard Forest de Belidor (1697-1761), who also wrote pioneering books on hydraulics and mechanics.[5] The son of a cavalry officer, in 1720 Belidor became professor of mathematics in the artillery school at Le Fere. This was a logical placement, for the Corps of Artillery was another source of engineering talent. It established its own training schools, thereby continuing the close association between the practice of gunnery and mathematical analysis that began with Tartaglia.

Among the French engineer-scientists of this era was Henri Pitot (1695-1771), the inventor of the pitot tube, a device that he used to measure the flow of streams. He could hardly have foreseen that his invention would eventually become indispensable in aviation, where it has long served as an airspeed indicator. Pitot was educated as a mathematician and astronomer, but became interested in the flow of water, eventually using his acquired knowledge while working on water supply and drainage projects, as well as maintaining the Canal du Midi.[6] In its transition from science to engineering, Pitot's career demonstrated the growing association of science and technology that became increasingly evident in the 18th century.

French engineers of the 18th century were a distinctive group, for their educational experiences diverged sharply from those of previous engineers. Most were academically trained, and many parlayed their training into a preeminent role in the founding of engineering science. They also differed from previous generations of European engineers in that their profession was consciously organized by agencies of the central government, specifically the army and the Ministry of Finance, the latter being the agency that created the *Corps des Ponts et Chaussees*.

The engineers for the most part seem to come from about the same general level of French society. Most of them were recruited from the ranks of the lower nobility and the professional ranks—lawyers, physicians, architects, and other engineers. Sons of army officers are also well-represented. There are not, as among Italian Renaissance or contemporaneous

British engineers, any prominent French engineers of humble origins. Under the *Ancien Regime* engineering was a respected profession, but not a route of upward mobility. Sabastien Vauban, a renowned designer of fortifications, became a Marshal of France just before his death, but no other French engineer of the 18th century received any comparable recognition. Not that Britain was any more generous: contemporary British engineers were well-known and admired, but after Myddleton and Vermuyden in the early 17th century there was a lapse of two hundred years before another British engineer was knighted, and none was raised to the peerage.[7]

Although 18th century French engineers owed the establishment of engineering as a distinct profession to the policies of the central government, they were by no means its passive servants. As social ferment spread across France, many engineers provided intellectual leadership for those who sought to reform or even overthrow the traditional order. Vauban at the beginning of the 18th century and Coulomb at its end urged such steps as the revision of the inequitable tax system and abolition of internal trade barriers, while Lazare Carnot and Monge were among the leaders of the Revolution.

Significantly, the overthrow of the monarchy was accompanied by an improved status for France's engineers. The Revolutionary government expanded France's already excellent system of scientific and technical education by creating the *Ecole Polytechnique*, with Monge as its first director and a curriculum that required three years of basic sciences and mathematics before the study of engineering was undertaken. No prominent engineer appears as a victim of the post-revolutionary Reign of Terror. The great chemist Antoine Lavoissier fell victim to the guillotine, and appeals to spare him because of his scientific brilliance were allegedly shrugged off with the comment that "the Republic has no need for savants." Yet the careers of Carnot and Monge indicate that Republican policies belied this opinion. Lavoissier was not executed because of an abhorrence of scientific elitism, but because he had worked as a tax collector, and because he had won the enmity of scientists less talented than he. The Revolution was not anti-scientific, and the early Republic expanded the roles of French engineers. And as we shall see, subsequent governments bolstered the position of French engineering as a whole.

The Civil Engineer in Britain

British civil engineering was initially nurtured by river improvement and drainage projects. In the 18th century it expanded rapidly beyond these origins, largely because Britain, like France, needed to improve its internal communications. In Britain, however, the pressure for better transport facilities came from the growth of commerce and industry; military considerations and governmental influence were secondary. The condition of British roads in the early part of the 18th century can only be described as wretched. Road building and maintenance were the responsibility of local parish authorities, even for supposedly main highways, and this system was pathetically ineffective. Roads were usually impassable in bad weather; the use of wheeled vehicles was very limited, and carriage of goods by land was usually done by pack animals, which meant small loads and high costs. Some efforts were begun in the late 17th century to improve matters by creating turnpike trusts. These were combinations of local governmental authorities (not, as in 19th century United States, private corporations) given the power to impose tolls, but by the mid-18th century little tangible progress had been made.

Inland water transport was similarly inadequate at the beginning of the 18th century. British canal building did not commence until after the middle of the century, and few British rivers are navigable for any appreciable distance inland. This was an intolerable situation for an economy that was based on a vigorous commerce and was beginning to industrialize. There was thus a powerful incentive to build roads and canals, and consequently a demand for people with the ability to build them.

The British response was in striking contrast to that of the French. The British government established no engineering colleges, nor was the subject taught at any British university until much later. British civil engineers continued to come from traditional sources: through apprenticeships to millwrights, masons, or other crafts, or simply through their own efforts at self-education. Yet despite massive governmental indifference to their cultivation, British engineers were held in high public regard. The greatest of the engineers became household names, and their lives and achievements were written about extensively, so that for the first time in history we have an almost overwhelming abundance of knowledge about engineers instead of a frustrating scarcity.

These emerging British civil engineers differed from their French counterparts not only in the lack of formal training for most of them, but also in the fact that they were drawn from practically all levels of British society. They were also more likely to be involved in mechanical matters, a reflection of the accelerating industrialization of Great Britain.

The first well-known British civil engineer of the 18th century came from the lower ranks of British society. He was James Brindley (1716-1772), a pioneer canal builder and son of a crofter in Derbyshire who was reputed to have been a very lackadaisical worker.[8] Brindley was educated by his mother and then apprenticed to a local wheelwright and millwright, a route followed by many young men who subsequently became engineers. When he set up in business his initial activities were largely mechanical, although he first attracted attention by building a tunnel at Irwell in the Pottery District to channel water for a wheel that pumped out a coal mine. For this feat he received two shillings a day. In 1754 the Duke of Bridgewater engaged Brindley to build a canal connecting the Duke's coal properties at Worsley with the fast-growing textile center of Manchester, beginning an association in canal construction that would last until Brindley's death. The Duke's interest in canal building may have been fired by his inspection of the Languedoc Canal while he was on a Grand Tour of Europe, at that time an obligatory junket for a young English nobleman.[9]

The first road-builder of note, John Metcalf (1717-1810) had an even less promising start. He was born to a poor family in Knaresborough, Yorkshire.[10] For a time he supported himself by playing the violin, and served as a musician in a volunteer regiment that served King George II during the Jacobite Rebellion of 1745. He then became a recruiting sergeant and participated in the Battle of Falkirk. Metcalf then turned his attention to road building, stimulated by an Act of Parliament that authorized the construction of a turnpike between Harrogate and Buroughbridge. His skills in laying out a road were noted by one contemporary: "I have several times met this man traversing the roads with the assistance of only a long staff, ascending steep and rugged heights, exploring valleys and investigating their several extents, forms, and situations, so as to answer his design in the best manner....I made some inquiries respecting this new road. It was really astonishing to hear with what accuracy he described its course and the nature of the different soils through which it was conducted."[11] Metcalf went on to construct an extensive network of roads in Yorkshire, and he was still at it, building roads in Lancashire while in his seventies. Perhaps none of his accomplishments were truly out of the ordinary, except for one thing: John Metcalf had been blind from the age of six.

While Metcalf had to struggle with poverty and physical handicap, another British engineer had a more fortunate beginning. John Smeaton (1724-1792), the first man to use the title "civil engineer," was the son of a prosperous attorney in Leeds. It was a severe blow to his family when he chose to become an engineer.[12] Engineering was still only marginal as a profession, and in snobbish Britain an occupation that could be entered by men like Brindley and Metcalf ranked low in the social hierarchy. More than anyone else, Smeaton changed this attitude.

It is illustrative of the status of engineering in Britain at the time (and in contrast to the contemporary French situation) that Smeaton's formal training consisted of apprenticeship to a maker of scientific instruments. His father regretfully accepted his son's choice and paid for his apprenticeship. He could well have afforded to send his son to an engineering college like those in France, but Britain had none.

Smeaton's engineering achievements were numerous and notable. His first and most dramatic was the successful construction of the Eddystone Lighthouse on a low, dangerous ledge a few miles offshore from Plymouth, where two previous structures had failed to survive violent storms. The result was that the Eddystone, heretofore a dreaded menace to shipping in the English channel, became a beacon of safety. Like many other early civil engineers, Smeaton also took a keen interest in what we would today call mechanical engineering. He constructed model water wheels and used them for a series of tests in order to determine the efficiency of different types—the first systematic, quantitiative experiments using scale models of actual machines.[13] He later made a number of improvements to the Newcomen steam engine that more than doubled its efficiency.

Smeaton's accomplishments as an engineer are a matter of record, but great as they were, they were not his only contribution to the growth of his profession. When he became established as a civil engineer, Smeaton initiated informal weekly meetings with others who shared the same occupational interests and who followed his example by calling themselves civil engineers. In due course this group became more formally organized and called itself the Society of Civil Engineers, less formally known as the Smeatonian Society. The formation of this group marked an important step in the development of engineering as a profession; engineers were beginning to think of themselves as a collective body with certain interests in common. Disagreement among the members caused the dissolution of the Society in 1792, but the period of its existence was an important phase in the separation of the engineer from the artisan.[14]

Smeaton himself and several other Smeatonians were members of one of the world's first scientific associations, the Royal Society, and they made many contributions to it. Among its members was John Grundy (1719-1783), the son of a self-trained engineer.[15] Grundy worked on the problem of Fen drainage, at the peak of his career receiving the substantial wage of one guinea per day plus all expenses.[16] Another was Thomas Yeoman, who also worked on Fenland drainage, built turnpike roads, became an expert on ventilation, and for a time served as president of the Smeatonian Society. In 1756 he became chief Marine Superintendent for the Royal Navy in charge of ventilation on Naval Vessels.

By the end of the 18th century the efforts and achievements of Smeaton and his associates had resulted in getting civil engineering recognized as an occupation with a status considerably higher than the crafts from which it had sprung. It was not yet a "learned profession," and in social status it was strictly middle class, but it was definitely moving ahead.

The mission of civil engineering was continued by men of still greater public stature. Two of the most highly acclaimed were Thomas Telford (1757-1834) and John Rennie (1761-1821). Both were Scotsmen, a point that illustrates another significant feature of 18th century British civil engineering. Brindley was born in the Peak District of northern Derbyshire, and Metcalf and Smeaton hailed from Yorkshire. Their associates came predominantly from the same general area or from Scotland. Most of them eventually gravitated towards London, which was understandable since it was the center of commerce and finance along with being the capital, but their origins were in the industrializing northern parts of Britain.

Telford and Rennie were both farmer's sons, but that was the extent of the similarity of their origins. Telford's father was a poor crofter who died when Thomas was a small boy. He was brought up in poverty by his mother, helped out by doing odd jobs, and eventually was apprenticed to a stonemason.[17] He was well-liked and evidently was recognized as a young man of promise, what his neighbors would have called "a laddie o' pairts." An uncle paid his fees at the parish school and a lady in the vicinity gave him use of her library. When he mastered his trade he went to London, became known as an expert mason, and through a program of vigorous self-education went on to a distinguished career in architecture and engineering.

By contrast, Rennie's father was a prosperous farmer in East Lothian, not far from Edinburgh. John Rennie was five years old when his father died, but he was not left penniless like Telford. His older brother George managed the family farm successfully and allowed young John to spend much of his time with a neighboring and quite renowned millwright named Andrew Meikle.[18] Rennie subsequently went to high school in Dunbar and then spent three years at Edinburgh

University, where he availed himself of the scientific curricula offered by John Robison and Joseph Black. The educational opportunities to be found in Scotland were far better than those in England, whose universities were still largely innocent of scientific instruction. Rennie thus had not only a thorough practical training, but a formal education in science that made him virtually unique among British engineers of the time.

Telford has been ranked as the greatest of the pre-railway British civil engineers. His achievements included bridges, canals, and above all, roads; his friend, the poet Robert Southey, dubbed him "the colossus of roads." His road-building techniques were very similar to those of his French contemporary, Tresauguet, and also to those introduced by a fellow Scot, John Loudon McAdam (1756-1831). The latter wrote extensively on improved methods of road construction, emphasizing, as did Telford and and Tresauguet, the critical importance of drainage, but he was primarily an administrator, and not himself a builder.[19]

Much of Telford's work was directed at improving communications in his native Scotland. His outstanding achievement in canal-building was the Caledonian Canal, which traversed the Great Glen across the Scottish Highlands. It was a project on which Telford pinned great hopes for bringing to the Highlands the economic and intellectual vitality evident in Lowland Scotland, but in that respect it proved a disappointment. Building the canal was more expensive and time-consuming than had been anticipated, and by the time it had been completed, changed conditions had reduced its usefulness. Outside Scotland, Telford distinguished himself by constructing a highway through the rugged mountain country of North Wales, and building a boldly conceived suspension bridge across the Menai Strait, connecting the road between England and Holyhead, the most direct route to Ireland.

Like Telford, John Rennie also made his way to London; one is reminded of Samuel Johnson's acerbic comment that "the fairest prospect any Scotsman ever sees is the road that leads to London." It was a natural progression; England was much the wealthiest part of the United kingdom, and for an able and enterprising young Scot it offered far greater opportunities than he could hope to find in his impoverished homeland. Rennie himself noted this fact in a ditty that he wrote in 1797, as far as is known his only venture into poetry:[20]

> Barren are Caledonia's hills
> Unfertile are her plains
> Barelegged are her brawyn nymphs,
> Bare arsed are her swains.

The muse seems to have deserted Rennie at this point, something for which we all can be thankful.

Rennie's initial destination in England was not London, but the Soho works of Boulton and Watt in Birmingham, which he visited in 1783 armed with a letter of introduction to James Watt from Dr. Robison. A year later, Watt, after sounding out Robison about his protegee, employed Rennie to supervise the construction of the Albion Mills in London, an ambitious project for a flour mill that was the largest application of steam power to factory operation attempted at the time.[21] Boulton and Watt had the contract for the machinery, but needed an engineer of demonstrated competence in mill work, and Rennie was chosen. He carried out his mission skillfully, but the mill was never a financial success. It was not a financial drain for long; it burned down three years after being put into operation.[22]

With Rennie, in fact, we come to a new phase in the evolution of the civil engineer, one that demonstrates the breadth of his abilities and responsibilities. Rennie died before the railway era began, but he was strongly interested in the development and use of the steam engine, and today would be considered a mechanical as well as a civil engineer. In regard to the customary activities of the latter, he worked on Fen drainage, built bridges and docks, and emulated Smeaton by building a lighthouse on the eastern coast of Scotland. Rennie's last great work was the construction of a new London Bridge that replaced Peter of Colechurch's 600-year old structure. Rennie died while building was still in progress, and it was finished by his son John, who, as we saw in the last chapter, was knighted for the achievement.[23]

Both the elder Rennie and Telford were concerned with elevating the status of engineers. One way of doing this was to strengthen associations devoted to the advancement of professional concerns. Rennie revived the Society of Civil Engineers shortly after Smeaton's withdrawal. Although it met weekly, it was a limited group dominated by the Rennies. In 1818 some young engineers decided that they needed a separate organization that would foster their own careers and their profession in general. The result was the formation in that year of the British Institution of Civil Engineers. After a slow start, the Institution came to life when Telford became its president in 1820. With his great prestige behind it, the Institution became securely established, and largely through Telford's influence it received a royal charter in 1828, another indication of British engineering's rise in status and prestige.

Britain and France Compared

Some of the differences surrounding the development of civil engineering in France and Britain have already been mentioned, but it is worth discussing them in greater detail. In France the *Corps du Genie* was a military unit that took on civilian projects, while the *Corps des Ponts et Chaussees* was the creation of the Ministry of Finance. In Britain, the Royal Engineers had nothing to do with non-military works, and the Institution of Civil Engineers was a wholly voluntary organization, with no relationship whatever to any governmental authority. The chartering of the Institution came on the initiative of its membership ten years after its founding, and the charter itself was drafted by its own secretary.

As noted above, early British civil engineers represented a broader cross section of their society than did their French contemporaries. In neither country do we find engineering regarded as an appropriate field for sons of upper-class families, but while French engineers were drawn from bourgeois families or sometimes from the minor nobility, British engineering included not only these classes, but also offered an avenue of opportunity for poor boys. Brindley, Metcalf, and Telford had no counterparts in France, in part a consequence of the strong tendency of British engineers to enter engineering through craft training as masons, millwrights, and the like. Only Rennie had an academic background comparable to what was customary for French engineers of his day. And even in his own family there was no set pattern; Rennie sent his younger son to the University of Edinburgh for the kind of scientific training he once had, but the son and his older brother learned their engineering in the family business.

Probably because of their predominantly on-the-job training, British engineers of this period made less use of scientific theories and data, and were less concerned with the theoretical aspects of engineering than the French. Brindley, Rennie, and Telford were all elected Fellows of the Royal Society and read papers before it, but they were not engineer-scientists at the level of Coulomb and Monge; indeed, no British 18th century engineer appears to fit in this category. Nor did the British write texts on engineering like those produced by Belidor or Pitot. The most conspicuous British treatises on engineering subjects were McAdam's works on road construction, but McAdam, as was noted earlier, was an administrator with no actual engineering training or background.

As a final contrast, British engineers displayed none of the political activism that appeared in France with Vauban, and made Carnot and Monge leaders of the Revolution. Admittedly, they had less incentive; discontent in 18th century Britain came nowhere near the intensity of feeling in France, and it was expressed mainly by literary intellectuals who had only the vaguest notions of how common people felt. The one exception that can be identified was Telford, who certainly had a first-hand knowledge of poverty. He was an enthusiastic reader of Thomas Paine's great revolutionary tract, *The Rights of Man*, and sent a copy to a friend in the Scottish town where he had served his apprenticeship. The pamphlet inflamed some of the town's young men to the point where they were jailed briefly for disturbing the peace, which consisted of drinking revolutionary toasts in the town square. In sending the work, Telford had used the postal privileges of his employer and patron, a member of the House of Commons, and for this he received a severe reprimand. After this incident he devoted his attention to engineering projects.

The Railroad Engineers

The role of the civil engineer was augmented and significantly transformed by one of the greatest of 19th century technologies, the railroad. Railroad building required all of the skills that the highway and canal engineers had, while at the same time requiring them to concern themselves with the mysteries of mechanical propulsion. Many engineers took on these varied tasks single-handedly, for civil and mechanical engineers were not distinct groups. Rennie and Telford were well aware of the possibilities of steam power, although the latter would have preferred to see steam applied to road locomotion. Rennie's son George helped to shape the new era as both a railroad engineer and a builder of locomotives and marine engines.[24]

The invention and development of the steam engine as a source of industrial power will be discussed in the following chapter. It is sufficient to note here that by the end of the eighteenth century the possibility of steam propulsion had been clearly recognized, and within the early years of the nineteenth it had become practical. The early development of steam locomotion was a largely British affair, but one Frenchman might have had a major role if given a little more encouragement. He was Nicholas-Joseph Cugnot (1725-1804), one of the many engineers about whom we know more of his works than his life. The son of farmers in Eastern France, Cugnot wrote on military engineering subjects, and in 1769 was a successful competitor in a project for a mechanical artillery vehicle sponsored by General Jean-Baptiste de Gribeauval, inspector of artillery in the French army.[25]

Cugnot's vehicle was a three-wheel contrivance with the engine over the front wheel and the boiler protruding out front. Not only did it represent the first application of steam power to land transport, its two-cylinder engine was the only high-pressure steam engine built in the 18th century.[26] It was a pioneering design, but it was awkwardly balanced and difficult to steer. The government soon lost interest and the vehicle became a museum piece. Cugnot has been credited with building the first automobile, but that is stretching things a bit, for his vehicle had no influence over the subsequent development of either the steam engine or the mechanically propelled road vehicle.

Steam locomotion was first successfully applied by Richard Trevithick (1771-1833) in Great Britain and Oliver Evans (1755-1819) in the United States. Both began with road vehicles using steam pressures of 30-50 pounds per square inch, and both exhausted the steam from the cylinders through the smokestack, thereby increasing the draft in the firebox.[27] Evans, a versatile genius, will be discussed in Chapter 7. Trevithick was born in Cornwall, the son of a mine manager. His formal education went through secondary school, after which he found employment as a mining engineer.[28] He lived close to William Murdock, who worked in the West Country for Boulton and Watt, and he probably knew Jonathan Hornblower, who experimented with a compound steam engine (see Chapter 6). Trevithick was certainly familiar with the use of steam power for pumping water out of mines, and Murdock may well have directed his interest to steam locomotion.

Trevithick built a steam carriage in 1801 and patented it a year later. In 1804 he put a steam locomotive into operation on a tramway in a Welsh ironworks and ran another on a mine tramway in Northern England a year later. These vehicles were unquestionably the world's first railroad locomotives, but they were too heavy and rigid (they had no springing) for tracks that originally had been laid for animal-drawn wagons. Trevithick gave up on his experiments and left for a mining enterprise in Peru that was to leave him impoverished when the property was wrecked during the Peruvian War of Independence.

Trevithick had the misfortune to be a bit ahead of his time. His ideas subsequently took hold, and in the early years of the 19th century crude steam locomotives began to appear in the collieries of northern England. The tramway, which used wooden or iron rails to facilitate the hauling of coal-laden carts by draft animals, had long been a fixture in the mining districts, and the substitution of steam for animal power was a natural progression.

This was the environment in which George Stephenson (1781-1848) grew up. Born near Newcastle, a name virtually synonymous with coal mining, his life and career exemplified how an engineering career could offer dramatic upward mobility. His father tended pumping engines, moving from job to job as the seams were exhausted and new pits were opened. George had to go to work as soon as he was able, had no schooling, and was illiterate until the age of eighteen.[29] Yet he displayed a genius in handling machinery, and he soon rose above his father's level and at thirty he was serving as engine wright of Killingsworth colliery. While he was there he invented a reliable safety lamp for miners, a feat usually overlooked because it was eclipsed by his later achievements, although at the time it involved him in an acrimonious controversy with the supporters of Sir Humphrey Davy over credit for the invention.[30] Actually, it was a clear case of simultaneous and independent invention in response to a pressing need.

At Killingsworth, Stephenson had ample opportunity to become acquainted with the crude steam locomotives that were going into service in the coal fields of northern England, and to appreciate both their current deficiencies and their future potential. He did not invent either the steam locomotive or the railroad; his genius lay in the fact that he saw beyond the hauling of coal carts to a revolutionary system of land transportation based on locomotives, cars, and tracks.

His opportunity came with a projected rail line to connect the coal fields around Darlington in Durham County with the seaport of Stockton. Stephenson's reputation was already strong enough for him to be given the responsibility for building the line—history's first common carrier railroad—and for him to persuade the directors to use steam power. The Stockton and Darlington Railway was opened for traffic in 1825. Within five years Stephenson was building a much more important line from Liverpool to Manchester. The locomotives as well as the line itself were Stephenson's work, assisted by his son Robert (1803-1859), then on the threshold of a career that would win him recognition as one of the greatest of 19th century British civil engineers. For his efforts, the Stockton and Darlington company paid George Stephenson £ 300 a year—not a rich reward for bringing the railroad age into being, but a fair distance from the twelve shillings a week he had made when he first started tending colliery engines.

Along with his other achievements, George Stephenson had the distinction of being made the first president of the British Institution of Mechanical Engineers, which was founded in 1847, a year before his death. Engineering was showing signs of specialization, even fragmentation, and this marked its first organizational division. The formation of the new society bore witness to the growing rivalry between the established road and canal engineers and the new generation of railroad engineers. There was also, as the next chapter describes, a growing body of engineers concerned predominantly with mechanical matters, so a split was all but inevitable.

George Stephenson was the classic example of the untutored genius, but his lack of schooling was not by choice, and he saw to it that his son received the education he had lacked. Robert Stephenson was sent to good schools, including a private school in Newcastle, attended lectures in science at Edinburgh University for six months, served an apprenticeship at Killingsworth colliery, and was drilled rigorously at home by his father.[31] His achievements are well known: numerous railroad lines in Britain and elsewhere, the boldly conceived tubular wrought iron bridge across the Menai Strait, the same basic design repeated on a larger scale across the St. Lawrence at Montreal, and other feats.

In planning the design of the Menai Bridge, Robert Stephenson was truly exploring uncharted territory. He first proposed to build a girder bridge comprising two spans of 460 feet each, a more than sevenfold increase over the longest girder bridge previously built.[32] He was determined to construct a bridge of iron, but lacked a grasp of the actual properties of wrought iron when subjected to the stresses that a bridge might encounter. He therefore conducted a set of experiments in a shipyard in which iron beams of different cross-section were loaded until failure occurred. The experiments were vindicated by the successful service of the bridge, but equally important, they provided a theoretical base for modern structural engineering.[33]

Robert Stephenson can be regarded as one of the greatest British civil engineers of the 19th century, yet in history he has been overshadowed by his father, in part because he was a modest and unassuming person who outlived his father by only eleven years. In contrast to his father, he was elected to the Institution of Civil Engineers and served as its president for two years just before his death.

With another of the great engineers the situation was reversed. Isambard Kingdom Brunel (1806-1859) was also the son of an eminent engineer, Sir Marc Isambard Brunel (whose career is described in the next chapter) but in this case the son outshone the father, and he was neither modest nor unassuming. Like Robert Stephenson, I.K. Brunel received excellent formal training. After attending an English boarding school, he was sent at the age of fourteen to the College of Caen and then to the Lycee Henri-Quatre in Paris, a reflection of his father's French origins.[34] He also served an apprenticeship with Louis Breguet, a famous maker of chronometers, watches, and scientific instruments, after which he returned to England to work with his father.

Brunel was a self-assertive genius whose achievements are even better known than Robert Stephenson's. He built the Great Western Railway, the largest British system during the period of private ownership, and persuaded the directors to use a seven-foot gauge for the tracks. It was a daring choice—too daring to win general acceptance. The "standard" gauge of 4 feet, 8 1/2 inches was a historical accident, employed by Stephenson simply because it had previously been used in the collieries. Although the wider gauge made for greater car loadings and smoother operation, it also resulted in higher construction costs. No other railroad followed suit, so that when the interchange of traffic and rolling stock became an essential consideration in railroad operation, the greater mileage of standard gauge track was decisive, and the Great Western converted to it after Brunel's death.

Not content to concentrate on railroads, Brunel engaged in heroic feats of ship design and construction. He designed the *Great Western*, whose voyage from Bristol to New York demonstrated the feasibility of steam-powered transatlantic crossings.[35] Far more famous was his *Great Eastern*, which at 18,000 gross tons and 32,000 tons displacement was for many years the world's biggest ship by a large measure. The *Great Eastern* demonstrated something that most engineers should always keep uppermost: technical virtuosity has to be matched by commercial practicality. The ship was meant for long voyages to the Far East and Australia, but it was too big for available harbor and dock facilities in that part of the world, and it therefore proved a commercial failure. It eventually was put to good use in the laying of the first transatlantic telegraph cable.

Brunel did not lack for self-confidence, and he was of the strong opinion that a qualified engineer should be left to go about his work as he saw fit. His thoughts on the subject of the relationship between an engineer and his employers leave little doubt regarding how authority should be distributed:[36]

> A board of directors has a perfect right to dispense with the services of an engineer, or to lay down any rules of conduct they may think fit, and to engineer the works themselves, either as a body or by appointing one of its own members, if they think fit; but if they desire to have the advice and responsibility of any respectable engineer—at all events, if they wish to have mine—they must place the usual amount of confidence in me; and as long as I am engineer they must leave me to conduct the engineering, and must act as if they assumed that I was more able to advise the Board upon all the usual practical questions of engineering than any one of the Directors. Admitting as I do the full right of a Board of Directors to determine whether they will have an engineer or not, if they do have one they must trust him to do his own work.

This is a clear manifesto for the engineer's autonomy. Brunel himself undoubtedly enjoyed a considerable freedom to pursue his vision of how engineering should be done. His career certainly contributed to the heroic stature of the Victorian engineer. It is not at all certain, however, that he was typical of contemporary engineers or those to come later. This is a topic to which we will return in subsequent chapters.

Isambard Kingdom Brunel and Robert Stephenson both died in 1859, neither of them even reaching their sixtieth year. How much more might they have achieved with even another ten years of life? It makes for interesting if rather melancholy speculation, but that is all it can be.

The third great civil engineer of the time, Joseph Locke (1805-1860) has not enjoyed the same measure of fame as Brunel and Stephenson, yet in his day he was considered fully their equal, and his life span almost exactly matched theirs. He was born in Sheffield, and like Stephenson he had mining connections, except that his status was considerably higher, for his father was the manager of a colliery. He was educated at Barnsley Grammar School and learned his engineering as an apprentice to George Stephenson.[37] His career was an international one; he built railways in Britain, France, Spain, and the Netherlands. In addition to his technical contributions, Locke was particularly noteworthy for his careful efforts to bring accurate accounting methods to railroad construction. He saw to it that the contractors who built the railroad were given precisely drawn up specifications of their tasks. At the same time, he closely supervised their performance.[38] Locke's career thus presaged an increasingly important role for the engineer, one that combined technical skill with the financial and managerial oversight of projects.

British railroad engineers were not the only ones involved in this new means of transportation, but they were the first, and for some time the most eminent. Their strongest competitors were in the United States, where railroads addressed a pressing need for improved inland transportation. The story of American engineers will be told in Chapter 7. On the European continent railroad development, like industrial development, lagged behind Britain, and in its early stages British engineers were much in demand. British engineers built Japan's first railroads, and today Japanese road and rail traffic still adheres to the left-hand rule. To an even greater extent than ever before, engineering was becoming an occupation of international scope, and one that was defined by, and at the same time defining, the emerging industrial era.

Notes

1. Esther C. Wright, "The Early Smeatonians," Newcomen Society *Transactions*, 18 (1937-38), p. 107.

2. Hans Straub, *A History of Civil Engineering* (Cambridge, Massachusetts: MIT Press, 1964), p. 20.

3. Maurice Daumas, *A History of Technology and Invention*, vol. 3 (New York: Crown, 1969), p. 245.

4. R.S. Kirby and P.C. Laurson, *The Early Years of Modern Civil Engineering* (New Haven: Yale University Press, 1952), p. 59, and J.P.M. Pannell, *An Illustrated History of Civil Engineering* (New York: Fred Ungar, 1965), p. 24.

5. James K. Finch, *The Story of Engineering* (Garden City, New York: Doubleday, 1960), p. 168.

6. Straub, op. cit., p. 129.

7. William Thomson (1824-1907) was elevated to the peerage as Lord Kelvin in part for his crucial role in the laying of the trans-Atlantic cable, but he was also honored for his seminal scientific work in electricity and thermodynamics.

8. Samuel Smiles, *Lives of the Engineers* (London: John Murray, 1904), p. 74. A good, up-to-date biography of Brindley is C.T.G. Boucher, *James Brindley, Engineer, 1716-1772* (Norwich: Gosse and Son, 1968).

9. L.T.C. Rolt, *From Sea to Sea: The Canal du Midi* (Athens: Ohio Universitry Press, 1973), p. 3.

10. Smiles, op. cit., p. 90.

11. W.H.G. Armytage, *A Social History of Engineering* (Boulder: Westview, 1976), pp. 83-84.

12. L.T.C. Rolt, *The Mechanicals: Progress of a Profession* (London: Heineman, 1967), p. 6.

13. Aubrey F. Burstall, *A History of Mechanical Engineering* (Cambridge, Massachusetts: MIT Press, 1965), pp. 243-44.

14. Wright, op. cit., p. 103.

15 Ibid.

16 Ibid., p. 104.

17 L.T.C. Rolt, *Thomas Telford* (London: Longmans, 1958), pp. 2-5.

18 C.T.G. Boucher, *John Rennie, 1761-1821: The Life and Work of a Great Engineer* (Manchester: Manchester University Press, 1963), pp. 1-5.

19 *Dictionary of National Biography*, ed. Leslie Stephen and Sidney Lee (Oxford: at the University Press, 1917) vol. 12, pp. 395-97.

20 Boucher, op. cit., p. 21.

21 *Dictionary of National Biography*, op. cit., vol. 16, p. 905.

22 Phyllis Deane, *The First Industrial Revolution*, ed. 2 (Cambridge: Cambridge University Press, 1979) p. 127.

23 *Dictionary of National Biography*, op. cit., vol. 16, pp. 906-907.

24 Boucher, op. cit., p. 96.

25 Daumas, op. cit., p. 262.

26 Eugene Ferguson, "Steam Transportation," in Melvin Kranzberg and Carroll E. Pursell (eds.), *Technology in Western Civilization*, vol. 1 (New York: Oxford University Press, 1967), pp. 291-92.

27 Burstall, op. cit., p. 267.

28 *Dictionary of National Biography*, op. cit., vol. 19, pp. 1140-44.

29 L.T.C. Rolt, *George and Robert Stephenson, The Railway Revolution* (London: Longmans, Green and Co., 1960)

30 Ibid., pp. 30-34.

31 Ibid., pp. 15-18.

32 L.T.C. Rolt, *Victorian Engineering* (Harmondsworth: Penguin, 1970), p. 28.

33 Ibid., p. 30.

34 L.T.C. Rolt, *Isambard Kingdom Brunel: A Biography* (London: Longmans, Green and Co., 1957), p. 17.

35 Ibid., pp. 196-97. Another British steamer, the *Sirius* arrived in New York a few hours ahead of the *Great Western*, but it was almost out of fuel and was too small for trans-Atlantic service.

36 Isambard Brunel, *The Life of Isambard Kingdom Brunel, Civil Engineer* (London: Longmans, Green and Co., 1870), p. 480.

37 *Dictionary of National Biography*, op. cit., vol. 12, pp. 37-38.

38 Rolt, *Victorian Engineering* op. cit. p. 38.

Chapter 6

The Mechanical Engineer and the Industrial Revolution

The Industrial Revolution in Britain was one of the most important yet most elusive events in history. There can be no denying that the process of industrialization that accelerated during the last quarter of the 18th century brought fundamental changes to British society, changes that were eventually felt all over the world. Yet for all its historical importance, the Industrial Revolution cannot be located in a precise time frame. The term "revolution" implies a rapid and dramatic break with the past, but most chronologies of the Industrial Revolution assign at least seventy years to its unfolding. The revolutionary changes that it brought required many years, for the rate of change was modest; from 1760 to 1820 British industrial output grew at the rate of 1.5 percent each year, not much more than the rate at which economy as a whole grew.[1]

While assigning precise dates to the Industrial Revolution is problematic, it is even more difficult to determine its causes. Its roots can be found in a mixture of agricultural progress, prior commercial expansion, a fairly open social structure, the availability of capital, an adequate supply of natural resources, changed cultural values, and the development of new technologies. An accurate, if not very satisfying, account of early British industrialization is that a confluence of separate factors interacted in mutually reinforcing ways to produce cumulative changes in what was produced and the way it was produced.[2]

None of this should obscure the reality or the significance of the Industrial Revolution. From approximately the last quarter of the eighteenth century and more or less steadily thereafter, the British economy became increasingly dependent on the products of industry. A world based on agricultural products and human labor was slowly receding. Increasing numbers of people found work as employees rather than as peasant farmers and artisans. An accelerating pace of technological change brought forth a spate of new products and processes, while at the same time effecting fundamental changes in social and economic relationships.

The central importance of technological change in the Industrial Revolution is unquestioned. One economic historian who has been careful to narrate and analyze the multiplicity of factors that contributed to the Industrial Revolution speaks for many when he notes that "...in the last resort, the decisive factor both in increasing the scale and in changing the methods and location of production was technology."[3] Still, this statement, while accurate, begs an important question. Technological change was obviously central to the Industrial Revolution, but what accounts for the great efflorescence of technology at this time? Was it the result of a unique outpouring of hitherto dormant genius? Was it a response to new demands? Was it an outgrowth of a scientific revolution that had been gaining momentum for a century? Economic historians have tended to sidestep this issue.[4] As one economic historian has admitted, "Technological progress, in its widest definition, has remained as the basis of the Industrial Revolution, and it still poses a challenge to economists whose understanding of it has thus far been limited."[5] We will not attempt to take up the matter in its entirety here, but by focussing on some engineers of the 18th and 19th centuries, we hope to provide some insights into the lives and work of people who did much to advance industrial technology, while at the same time noting how a changing social and economic milieu stimulated their efforts.

The Industrial Revolution was unquestionably the era of the engineer, both for what he accomplished and for what he symbolized.[6] No longer viewed as simply a talented artisan or an artist who did some inventing on the side, the engineer was taking on a distinct occupational status. More than this, he was often seen as a heroic figure whose efforts were transforming the world. The most prominent of the engineers of this era were given widespread recognition in the popular media, most notably in the widely read works of Samuel Smiles. In a characteristic eulogy, Smiles said of them: "Our engineers may be regarded in some measure as the makers of modern civilization... Are not the men who have made the motive power of the country, and immensely increased its productive strength, the men who above all others have tended to make the country what it is?"[7]

We have already briefly examined the work of civil engineers; this chapter will concentrate on mechanical engineers, although the line between the two was often blurred until well into the nineteenth century. This chapter concentrates on British engineers, not because their contemporaries in other lands lacked in skills and achievements, but simply because Britain was the pre-eminent industrial nation until well into the nineteenth century. British engineering accomplishments were a model and an inspiration to the rest of the world. Before the century was over other countries would produce engineers of equal, and in some cases superior, talent. The stories of some of them will be told in subsequent chapters, but first we will concentrate on some of the men who made major contributions to Britain's pre-eminence as an industrial nation.

We have also restricted our narrative to engineers who worked in only two sectors of the emerging industrial economy: steam engines and machine tools. It is not that other industries were stagnant; throughout the Industrial Revolution crucial improvements were wrought in the production of textiles, iron and steel, chemicals, and numerous other products. All of these industries benefited from the application of engineering talent, and a complete narrative of engineering in the Industrial Revolution would have to include their stories. We have restricted our discussion to the designers of steam engines and machine tools in order to keep the chapter from being too lengthy, but the narrowed focus has not been completely arbitrary. These two industries were of central importance to the era; without denying the crucial roles played by other industries, we can simply note that steam engines and machine tools were indispensable for maintaining the momentum of the Industrial Revolution.

Some narratives of the Industrial Revolution focus on a number of dramatic inventions that are attributed to untutored geniuses employing unsophisticated cut-and-try methods. There certainly were enough examples to give a surface plausibility to this view. The great majority of inventors and engineers in the early phases of the Industrial Revolution had little in the way of formal education. However, few of them could be categorized as genuinely untutored. Most had trained to be craftsmen such as millwrights and instrument makers, and most of them had tackled self-education with considerable fervor. With few exceptions the early mechanical engineers had a good working knowledge of arithmetic, geometry, and theoretical as well as practical mechanics.[8]

In Britain the start of industrialization was energized by a sizeable cadre of individuals making technical contributions both great and small. The problem for us is to determine which of them can be properly classed as engineers. Conventional accounts of the Industrial Revolution have stressed epochal inventions such as the Bessemer converter and the power loom, inventions that are usually discussed in conjunction with the lives of their inventors, hence the emphasis on men such as Hargreaves, Crompton, Arkwright, and Cartwright. But not all inventors can properly be termed engineers, even though an engineer can be an inventor. The key difference is that invention is a one-shot affair, even though it might take a lifetime to perfect a successful device or process. In contrast, to be an engineer implies a long-term career. We will generally limit our discussion to those men who were more or less continuously occupied with engineering, and whose efforts ranged over a variety of technical tasks. On occasion, however, it will be necessary to mention the life and works of people who were more inventors than engineers, especially those who lived and worked at a time before engineering had been established as a regular career.

The Introduction of Steam Power

Although the steam engine was not the typical source of power during the early decades of the Industrial Revolution, it was nonetheless highly significant for the development of engineering. The creation and application of the steam engine posed a host of novel technical challenges. Steam power required greater skill and precision than did wind- and water-powered devices. Technical breakthroughs necessitated the solution of unique problems, many of them unforeseen. The steam engine also made heavy demands on the ability to think logically and systematically while at the same time exercising a good deal of creativity.

The steam engine was in part an unforeseen spinoff from earlier studies of atmospheric pressure conducted by Torricelli, Pascal, and Von Guericke. As we saw in Chapter 2, the idea of using steam to produce motion is an ancient one that goes back to Hero of Alexandria's *Aeolipile*, but the idea lay dormant until the seventeenth century. At the end of that century Denis Papin, a Huguenot refugee in England who had been pupil of Christian Huygens, hit on the concept of filling a cylinder with steam, and then creating a partial vacuum by condensing the steam, causing atmospheric pressure to move the piston. This was just a laboratory demonstration, and Papin, who had earlier invented the pressure cooker, decided that his idea could not be made to work on a large scale.

His work was undoubtedly known by Thomas Savery (c. 1650-1715), for both were members of the Royal Society. Although little is known of Savery's life, he seems to have come from a prosperous family of landowners and merchants in Devon, in the west of England.[9] He might have served as a military engineer; more likely he did engineering work in the tin mines of nearby Cornwall. These mines had been worked since ancient times, and were experiencing difficulties with water as the shafts went deeper. Coal mining in other parts of Great Britain was encountering a similar problem.

Savery's solution was a system whereby steam was introduced into a closed vessel; the steam was then condensed, which caused water to be drawn up from below and into the evacuated vessel, whereupon it was forced out by a new injection of steam. The process then began anew. The idea was ingenious, and the process required no moving parts except for some valves for admitting, exhausting, and directing steam.

On the other hand, the Savery device required steam at more than atmospheric pressure, something for which contemporary metal-working techniques were inadequate. Also, the constant heating and cooling of the container gave it a low thermal efficiency—less than one half of one percent.[10] Even so, it was superior to other available pumping techniques, and Savery engines were used in some numbers in mining areas.

Savery's work was soon superseded by his fellow Devonian, Thomas Newcomen (1663-1729), a great engineer whose genius has only recently come to be fully appreciated. Newcomen came from a family that had some claim to distinguished ancestry, although his father was a merchant of only modest standing in the town of Dartmouth. By the time Thomas was born his family had left the Church of England to be Baptists, and he remained a devout member of the Baptist Church throughout his life. Newcomen was thus a representative of a significant number of engineers whose religious background lay outside the established Church of England.

Newcomen's youth is obscure. He probably received a good basic education and he may have been apprenticed to an ironmonger in neighboring Exeter, "ironmonger" at that time meaning one who made as well as sold metal products.[11] He subsequently went into business as an ironmonger in Dartmouth about 1685, and in the course of selling his products became familiar with the Cornish mines and the problem of pumping water out of them. The time he spent at the mines must have introduced him to Savery's engine. The first Newcomen engine was built in the Staffordshire coal district in 1712, but there is evidence of earlier and apparently unsuccessful installations in Cornwall, presumably part of a long process of experimentation by Newcomen and his associate Joseph Calley (or Cawley). The latter was another member of the Baptist fellowship. He was a plumber, a skill that certainly would have contributed to the design of the engine.[12]

Like many other great discoveries, Newcomen's basic concept was simple; the difficulty—as is often the case with other great discoveries—was in the execution. In its essentials the engine consisted of a vertical cylinder, open at the top, with the piston connected to an overhead beam, the other end of which operated the pumping mechanism. When the piston was at the top of its stroke, the cylinder was filled with steam, which was then condensed by injecting water. Atmospheric pressure forced the piston down to provide the power stroke, and the process was repeated. Ten to twelve strokes per minute was the normal pace of a Newcomen engine, and it had the great advantage for its day of requiring steam only at atmospheric pressure. Calley seems to have been responsible for introducing the mechanism that operated the valve automatically by the action of the engine.

Crude and inefficient as it was by later standards, the Newcomen engine was a marked advance as a source of power. It also should be noted that the overhead beam for transmitting the power of the engine was a completely novel device. By 1781 about 360 Newcomen engines had been built,[13] mostly for pumping water out of mines, although a few may have been used as factory power plants. Some were still in operation in the early years of the 19th century.

Newcomen did not receive all of the rewards he was entitled to. He had to share his success with a group of speculators that had bought Savery's patents, which covered all engines for "raising water by the impellent force of fire."[14] Very few patents were issued in eighteenth-century Britain, and the procedures for granting and administrating them could be capricious. It was a flawed system, and although Newcomen might have been the first, he was certainly not the last engineer to be victimized by patent laws.

Thomas Newcomen's "atmospheric engine" was a crucial stage in the development of power sources independent of the force of wind, water, and human and animal muscle. Although subsequent refinements, notably by John Smeaton (see Chapter 5) made it more efficient, it could never have been adopted to the variety of industrial uses that would employ steam. For this to happen, fundamental improvements had to made; these came through the inspired efforts of James Watt (1736-1819).

By the time James Watt was born vital statistics were being kept with some accuracy in Britain, and his life has been so thoroughly studied that all of the basic facts are well known.[15] He was born in Greenock, Scotland, the sixth child (but the first to survive infancy) of a carpenter and shipwright, who also served as a town official. James was a sickly child, so his first schooling was done at home by his mother. When he attended the local grammar school his experiences seem to have been unhappy ones and his record was undistinguished. From his earliest years, however, he showed a liking and aptitude for mathematics, and he spent many hours in a home workshop.

When he finished school he went to work for his father, although not as an apprentice. Watt is unusual among his contemporary British engineers in that he never went through the conventional process of apprenticeship, although he certainly had equivalent training. In 1754 he decided to become a maker of scientific instruments and went to Glasgow to learn the trade. There, through a relative of his mother he met Robert Dick, Professor of Natural Philosophy, the first of the acquaintances at Glasgow University who would markedly influence his career. Dick convinced the young man that to become a first-class instrument maker he had to go to London, so in 1755 Watt journeyed to the capital with a letter of introduction from Professor Dick that helped him reach an arrangement whereby he would receive a year's instruction in instrument making for a fee of twenty guineas. This fee and his subsistence costs (coming to six shillings a week) were paid by his father, even though business reverses had left the elder Watt in difficult financial circumstances.

When his year of training ended, Watt returned to Scotland and resumed his acquaintance with Professor Dick, with the result that he was given the right to set up shop in the university district with the title "Mathematical Instrument Maker to the University." What is significant here is that Glasgow University was more than a place for instruction; it was also a venue for scientific research, some of it relevant to practical concerns. The connection with Glasgow University was fundamental to Watt's career. He came to know Joseph Black, who was conducting pioneering inquiries into the nature of heat, and John Robison, at that time a student at the university. Through these associations he acquired a useful background of scientific knowledge. More than this, being at Glasgow put him in the midst of the remarkable blossoming of intellectual life that was taking place in Scotland, an environment that nurtured Adam Smith in Economics, David Hume in Philosophy, and Robert Burns in poetry.

During the 1763-64 session a professor of natural philosophy asked Watt to repair a model Newcomen engine that was being used for demonstration in classes. Watt did so, and if he had been an ordinary craftsman the matter would have ended then and there. But James Watt was no ordinary craftsman. He combined the scientist's desire to know for the sake of knowing with the engineer's penchant for applying his knowledge to the solution of practical problems. He and Robison had done some experimentation with steam engines, but the main consequence had been to give Watt a lifetime conviction that the use of steam at higher than atmospheric pressures was impractical and dangerous—an attitude justifiable enough, given the limitations of contemporary metallurgical theory and practice. As he studied the engine he observed that the boiler ran out of steam after only a few strokes of the piston. Part of the problem was inherent in scaling down the full-sized Newcomen engine; when a cylinder is reduced in size the ratio of surface area to total volume increases. This meant that for the model a proportionately larger area produced more rapid cooling and the steam condensed too quickly.

More importantly, Watt came to understand that a Newcomen engine of any size suffered from a waste of heat due to the constant cooling and reheating of the cylinder. As so often happens, the solution came to him in a moment of relaxation, while he was strolling on Glasgow Green on a Sunday afternoon (walking in the park being one of the few activities acceptable on a Scottish Sunday in those days, and indeed until some time into our own century). There he had the sudden insight that the steam should be cooled in a condenser separate from the cylinder.

From this insight Watt moved step by step to the reciprocating steam engine. Along the way he produced such fundamental innovations as the double-acting engine, parallel linkage for the transmission of motion, and the use of the flyball governor to regulate the speed of the engine. Yet for all his inventive genius, Watt would not have been nearly so successful had it not been for the commercial acumen provided by two businessmen who went into partnership with him, John Roebuck and Matthew Boulton. As Watt later said of Roebuck, "To his friendly encouragement, to his partiality for scientific improvements and to his ready application of them to the processes of art, to his intimate knowledge of business and manufactures, and to his extended views and liberal spirit of enterprise, must in great measure be ascribed whatever success may have attended my exertions."[16] Sadly, Roebuck suffered some business setbacks that led to bankruptcy; the partnership ended and Watt tried to make ends meet by working as a surveyor.

One of Roebuck's creditors was Matthew Boulton, who forgave the debt in return for taking over the partnership with Watt. Boulton's factory produced a variety of metal products, and he was eager to find a better source of powering the machines that made them. Beyond this, Boulton had the vision to realize the great potential of Watt's engine. Sales of the engine did not show a profit for a number of years, and the enterprise was supported by the earnings of Boulton's other business interests, as well as part of his own fortune. More than ten years elapsed before Boulton and Watt earned any profits on the steam engine.

Watt also benefited from advances in techniques for working with metals, most importantly a device for boring cylinders patented in 1774 by an English ironmaster named John Wilkinson.[17] The fit between cylinder and piston was obviously crucial to the operation of the engine, yet when Smeaton was building improved versions of the Newcomen engine he had to make do with variations in the bore of a 28-inch diameter cylinder that amounted to the thickness of a little finger.[18] As a result of Wilkinson's improved methods, the 72-inch diameter cylinder of a Watt engine would deviate by no more than than the thickness of a sixpence, about .05 in.[19] The improved method of producing cylinders was a dramatic example of the mutual stimulation of steam power and improved metal-working, a topic to which we will soon return.

Watt's engine also benefited from progress in a more abstract realm. Although it has been said that science owes more to the steam engine than the steam engine owes to science, it is nonetheless true that Watt's engineering skills were augmented by the science of the times. It is not true, as is sometimes asserted, that Watt's invention of the separate condenser was prompted by Joseph Black's theory of latent heat,[20] but if Watt's engine owed nothing to formal scientific theories, its development was surely stimulated by the scientific atmosphere in which Watt was enveloped. Whether or not an engineer was in full possession of the latest scientific knowledge was not crucial; what really mattered was having a scientific attitude that gave full rein to experimentation, observation, and testing.[21] This way of thinking and working is the real legacy that Black bequeathed to Watt, for as Watt duly noted, "...the knowledge upon various subjects which [Black] was pleased to communicate to me, and the correct modes of reasoning and of making experiments of which he set me the example, certainly conduced very much to facilitate the progress of my inventions."[22]

Boulton and Watt were eager participants in the advance of science that was occurring toward the end of the eighteenth century. Both were members of the Lunar Society of Birmingham, a small group of men which also included such figures as Erasmus Darwin, Josiah Wedgwood, and Joseph Priestley. The Society was not a formal research body, but it bound together scientifically inclined men who could draw upon one another for advice and support as they pursued their manifold interests. In the words of the Society's modern historian, "They met in an atmosphere of mutual congratulation and exposition, bringing visitors, showing experiments, discussing problems."[23] Under its auspices "practical" men such as Boulton and Watt could draw on the knowledge of scientific investigators such as Priestley. The Lunar Society was not unique; similar groups could be found in other provincial towns.[24] All of them contributed in some way to the advance of industrialization. The connections between science and technology that they helped to forge may have been indirect, but they were of great importance in expanding technology's scientific dimension.

The conjunction of scientific interest and a concern with practical engineering was characteristic of the late 18th century. Many scientists were interested in industrial applications, while at the same time, many entrepreneurs and engineers had scientific interests.[25] Had Watt been able to devote himself to science he could have achieved a measure of distinction in that realm. He worked out the formula for calculating horsepower that is still in use, and he discovered the chemical composition of water shortly after Henry Cavendish had done so.

Newcomen and Watt were the dominant figures in the early development of the steam engine, but by no means were they the only ones. A father-and-son team of engineers from the West Country of England, Jonathan Hornblower (1717-1780) and Jonathan Carter Hornblower (1753-1815) took the design of steam engines in a new direction.[26] The elder Hornblower was the son of an engineer who had worked in the Cornish mines. Cornwall was an important site for Newcomen engines, and Hornblower built several of them. He and his son subsequently did the same thing for Boulton and Watt, and this led the younger Hornblower to design the first known compound engine, in which steam exhausted from one cylinder was used in a second one. But once again the British patent system caused difficulties, as Hornblower's engine was held to be an infringement on Watt's patent. The compound engine was successfully introduced in 1804 after Watt's patents had expired by a former associate of the younger Hornblower, Arthur Woolf (1766-1837).

Many British engineers learned their trade while working around steam engines. Several noteworthy engineers developed their skills while working at Boulton and Watt's firm. The most talented of these was William Murdock (or Murdoch) (1754-1839). The son of a Scottish miller and millwright, he was apprenticed in these crafts and subsequently went to seek a job with Boulton and Watt. He probably was fortunate that Watt was away when he arrived and Boulton was the one to hire him, for Watt considered English craftsmen to be more skillful than Scottish ones.[27]

Murdock was sent to Cornwall to supervise the erection of Boulton and Watt engines, at which he proved very successful. He made a number of improvements that bothered Watt, because Watt, while affable in his personal relations, tended to be jealous in matters touching on his professional life. Murdock was also the co-inventor of gas lighting, which has been characterized as "one of the most neglected of the 'great inventions'."[28]

Murdock also wanted to apply steam power to locomotion. He succeeded in building a model steam carriage, but Watt ordered a halt to this experiment, partly because of his aversion to high-pressure steam. This was not a blind prejudice; boiler explosions and other mechanical failures were a very real possibility even though high-pressure steam in those days did not exceed 50 pounds per square inch. The future development of the steam engine depended on the use of much higher pressures, but its safe use required vast improvements in producing and working metal. Improvements in the latter were duly wrought by a succession of mechanical engineers, whose careers we will now consider.

The Machine Makers

The successful development of the steam engine was in part the result of James Watt's exposure to the scientific climate of Edinburgh. Its commercial production owed much to the business skills of Matthew Boulton. There was also a third factor that was no less crucial to the successful engineering of the steam engine: the development of improved tools and procedures for working metal components to finer tolerances than had been achieved before. In turn, the steam engine was an important stimulus for these developments.[29] The pioneering mechanical inventions of the late eighteenth and early nineteenth centuries, of which the steam engine is notable but not exceptional, created a demand for tools that could work metal with much greater accuracy than previously. In regard to the steam engine, Smeaton had complained that "neither the tools nor the workmen existed that could manufacture so complex a machine with sufficient precision."[30] Watt's engine languished for five years after its invention because it was not possible to effect a steam-tight fit between the piston and cylinder.[31] Once developed, these new tools made it possible to build new and more complex mechanisms. Their design and construction was the work of a group of highly skilled workers and designers who in due course began to call themselves mechanical engineers.

The process began with numerous unidentified craftsman; the earliest name to be really prominent was Joseph Bramah (1749-1814). The son of a Yorkshire farmer, he was apprenticed to a carpenter. After he completed his training he moved to London, where he eventually set up a machine shop. Bramah is credited with the invention of the hydraulic press and with opening the way for further advances in the practical application of hydraulics. He also has a claim to be the inventor of the flush toilet through his development of an effective means of valve operation.

Sometimes an engineer's greatest accomplishments can be found not in inert mechanisms, but in the people whose lives and careers they influence. Overshadowing all of Bramah's accomplishments as an engineer was his service as a mentor to Henry Maudsley (1771-1831), who can be fairly credited with being the father of the modern machine tool. Maudsley was virtually unique among the British engineers of his day in that he was a Londoner by birth. His father had served the Royal Artillery, and after being wounded and invalided out, he became an artificer in the Woolwich Arsenal. Young Henry joined him there at the age of twelve as a "powder monkey" who filled cartridges. From this he progressed to working with wood and metal.[32] There he exhibited a great talent for working with metal. Unlike many of today's engineers, he had a craftsman's touch with the materials he used; long after he was established as a successful engineer, one of his employees noted that "It was pleasing to see him handle a tool of any kind, but he was quite splendid with an 18-inch file."[33]

Because of his evident abilities, he was recommended to Bramah when the latter needed someone to make tools and fixtures for the accurate manufacture of a lock he had invented. Their association lasted until 1797, by which time Maudsley had become the manager of Bramah's works. Still, he considered himself underpaid and left to go into business for himself.

Maudsley was responsible for making the lathe an essential tool for the production of precision machine parts from iron and steel.[34] The lathe is an ancient device, and it occasionally attracted the attentions of engineers, Leonardo da Vinci being one of the most noteworthy. Maudsley is usually credited with the invention of the slide rest (which holds the cutting tool) but this is not strictly accurate, for such a device had already been used by French watch and clock makers. Maudsley's accomplishment consisted in making the lathe and its component parts suitable for heavy machining. To this end he built the first all-metal lathes, whose rigidity insured a much higher level of accuracy. Accuracy was also served by ensuring that the spindles and dead-centers were perfectly aligned, and that the slide rests moved over perfectly planed surfaces.[35] In addition, he painstakingly constructed a lead screw that allowed a lathe to turn accurate screws of uniform pitch. In general, he pursued accuracy with great vigor; by 1830 he had constructed a micrometer accurate to .0001 in.[36]

One of the most significant contributions made by 19th century mechanical engineers lay in their creation of the equipment and processes that made mass production possible. Maudsley participated in one of the earliest attempts to make large numbers of standard products through the use of machinery. In this case, the products were pulley blocks for rigging sailing vessels. In cooperation with Sir Samuel Bentham, Inspector of Naval Works, and brother of the philosopher Jeremy Bentham, Maudsley constructed the machinery that had been invented by Marc Brunel. This in itself was a significant innovation; as Abbott Usher has noted, the venture marked "the beginnings of specialized engineering work in which the contracting firm executes plans furnished by the inventors."[37]

We will return to this enterprise shortly; we should first provide some biographical information on the man who conceptualized the system. Marc Brunel (1769-1849) was briefly mentioned in the previous chapter as the father of Isambard Kingdom Brunel, but he deserves recognition as a great engineer in his own right. He was born in France, the son of a tenant farmer whose family had occupied the same property for over three hundred years.[38] Brunel had an eventful and varied life; some episodes in it would make a good movie script. He intended to be a naval officer, and was tutored by a retired sea captain, Francois Carpentier, who also served as American consul in Rouen. By the time he received his commission and returned from his first trip at sea, the French Revolution was in full swing. Young, temperamental, and a royalist, Brunel got into an argument in a tavern, and was forced to take refuge with his former tutor. In his capacity as American consul, Carpentier was able to to provide Brunel with documents that enabled him to board an American ship and make his way to New York. There he engaged in surveying and eventually rose to the position of chief engineer for New York City.

After seven years Brunel journeyed to Britain with some mechanical ideas he hoped to market, among them the system for mass-producing pulley blocks. Mass production of a sort antedated the industrial revolution; Gutenberg's use of movable type for printing is, after all, an example of mass production. But Brunel's was the first successful effort to use precision tools, powered by steam, to produce a uniform product. It also marked one of the first applications of the circular saw and the cone clutch.[39]

The product itself was a simple one, but large numbers were vital to the British Navy for the rigging of a sailing vessel; a single seventy-four gun ship of the line needed some 1400 of them. The system was installed in the Portsmouth Navy Yard over a period of six years, and went into operation in 1808. It consisted of 44 machines, powered by a single 30 h.p. steam engine. This new technology heralded fundamental changes in manufacture, for aided by the new machinery 10 unskilled men could do the work of 110 skilled craftsmen using traditional methods. In the peak year the plant turned out over 130,000 blocks.[40]

Brunel received a nice income for his efforts. While the installation was in progress he received a guinea a day, and ten shillings a day when he was in Portsmouth. More importantly, he was to receive a sum equivalent to one year's savings in costs when the machinery was in operation. This brought him £ 17,000 in 1809.[41]

Brunel had a similar plan for making boots for the army, but just as he was getting started the Napoleonic Wars came to an end. Like many other engineers whose work was too closely tied to the defense sector he lost heavily when the contract was cancelled. In his subsequent career, his outstanding achievement was to build the Thames Tunnel, the first tunnel under a major river, using a tunneling shield of his own design that was inspired by his observation of the Teredo shipworm boring into wood.[42] The tunnel was begun in 1825 and finally opened in 1843, just before his death. Financial limitations prevented the building of a spiral carriageway leading to it, so the tunnel was initially limited to pedestrian traffic until the East London Railway began to use it in 1865.[43]

Henry Maudsley, Brunel's associate in the pulley-block enterprise, is an important figure in the history of mechanical engineering not only for his technical accomplishments, but also for his role as a teacher of other engineers. Early nineteenth century England lacked formal facilities for the training of engineers, and an apprenticeship to a working engineer was the most likely means of gaining the necessary skills. Maudsley counted among his pupils three of the greatest 19th century engineers: Richard Roberts (1786-1865), James Nasmyth (1808-1890), and Joseph Whitworth (1803-1887). Roberts, the son of a Welsh shoemaker, started as a quarryman and then became a pattern maker for John Wilkinson.[44] His association with Maudsley was accidental. When his name was drawn for militia service in his home county, he left and eventually made his way to London, where he found work with Maudsley.[45] After a few years he set up his own business in Manchester, where he built one of the first planing machines, a gear-cutting machine, and a slotting machine that drew inspiration from one of Brunel's mortising machines.[46] He also made some important contributions to the industrial lathe, and developed workshop techniques that brought a degree of standardization to the manufacture of locomotive parts.[47] Finally, he used a kind of punch-card system for the multiple drilling of rivet holes. With considerable justification he has been called the greatest mechanical inventor of the nineteenth century. Sadly, his single-minded pursuit of new inventions often overcame commercial considerations, and his life ended in poverty.

Nasmyth, the inventor of the steam hammer, was a Scot with an unusual background for a British engineer. He was born in Edinburgh, the son of Alexander Nasmyth, an artist of some distinction. He went to Edinburgh High School, but was uninspired by the teaching, which he later characterized as "a mere matter of rote and cram."[48] He learned much more from using his father's workshop and spending his spare time at a local foundry. After high school he went to the newly-founded Edinburgh School of Arts, which despite its name was a pioneer technical college that offered instruction to working men and mechanics in the scientific principles underlying their various occupations.[49]

While in school Nasmyth earned some money by making models of machines. From these he progressed to full-sized machines, including steam engines and steam carriages. In the course of these activities he decided that for the full development of his skills he needed to work under Maudsley, and in 1829 he set out for London. He showed Maudsley some of his models and drawings, and impressed him sufficiently to be given a place as Maudsley's assistant, without the need for going through an apprenticeship. He stayed with Maudsley until the latter's death in 1831, and then decided to go into business for himself, eventually settling in Manchester, where, he believed, the skills of craftsmen were superior to those in London.

His most famous invention, the steam hammer, came as a response to an unmet need; a wrought iron shaft was needed for the paddle of the steamship *Great Britain*, but none of the existing forging equipment was up to the task. Nasmyth's solution was an adaptation of the steam engine. A hammer-block was attached to the piston rod of an inverted steam cylinder. Then, as Nasmyth later narrated,[50]

> All that was then required to produce a most efficient hammer was simply to admit steam of sufficient pressure into the cylinder, so as to act on the under-side of the piston, and thus to raise the hammer-block attached to the end of the piston-rod. By a very simple arrangement of a slide valve, under the control of an attendant, the steam was allowed to escape and thus permit the massive block of iron rapidly to descend by its own gravity upon the work then on the anvil.

Nasmyth's invention of the steam hammer came as a response to a particular technical problem. But equally important, his invention embodied the dark side of the Industrial Revolution. Nasmyth was convinced that the great obstacle to industrial progress was the dilatory and obstructionist character of industrial labor. His answer to the problem was the steady development of new devices that would remove labor's autonomy through the removal of the workers' skills:[51]

> The great feature of our modern mechanical improvement has been the introduction of self-acting tools. All that a mechanic has to do, and which any lad is able to do, is, not to labour, but to watch the beautiful functions of the machine. All that class of men, who depended upon mere dexterity [sic], are set aside altogether. I had four boys to one mechanic. By these mechanical contrivances I reduced the number of men in my employ, 1,500 hands, fully one half. The result was that my profits were much increased.

The coupling of mechanization with the de-skilling of labor may have solved some of the immediate problems of industrialization. Over the long haul, however, it is likely that it has been counterproductive. It is now evident that the economic strength of a country is derived not from its machinery but from the quality of its workforce. In ardently espousing the replacement of workers' skills by sophisticated machinery, Nasmyth was contributing to a managerial culture that pitted labor against capital to the ultimate detriment of both.

The third of Maudsley's distinguished pupils was Joseph Whitworth. He was born in Stockport, outside Manchester, the son of a Congregational minister and schoolmaster. He attended an academy in Leeds until the age of fourteen, when he was placed with an uncle who had a cotton-spinning business in Derbyshire. But he had no interest in cotton manufacturing, and at eighteen he left for Manchester where he worked in a machine shop.[52] He went to London in 1825 to enter Maudsley's shop and eight years later returned to Manchester to establish his own business.

Whitworth's major contributions to engineering were a dedication to uniform standards for screw threads and an insistence on accuracy in measurement. As the historian of the company into which his was eventually absorbed notes, "Whitworth understood, as no one else had yet done, the importance of extreme accuracy in engineering in the Age of Steel, and his machine for measuring to one two-millionth part of an inch is perhaps his finest monument."[53]

The Crimean War put Whitworth into the manufacture of weapons, and his company became an important producer of armaments. He was also a strong advocate of improved technical education in Britain, even to the extent of suggesting what must have seemed very far-fetched in mid-19th century England: that some day the cotton trade might decline and Manchester therefore should be prepared to diversify its industries.

Another British toolmaker, William Fairbairn (1789-1874), contributed to precision by developing standard formulae for the strength of boilers, tubing, shafting, and so on. The son of a Scottish agricultural worker who eventually became a farm manager, he learned much of his engineering while apprenticed to a millwright.[54] He has an especially interesting comment about the role of the millwright in the evolution of mechanical engineering:[55]

> In those days a good millwright was a man of large resources; he was generally well-educated, and could draw out his own designs and work at the lathe; he had a knowledge of mill machinery, pumps, and cranes, could turn his hand to the bench or the forge with equal adroitness and facility. If hard pressed, as was frequently the case in country places far from towns, he could devise for himself expedients which enabled him to meet special requirements, and to complete his work without assistance. This was the class of men with whom I associated in my early life—proud of their calling, fertile in resources, and aware of their value in a country where the industrial arts were rapidly developing.

The Social Origins of British Engineers

History is full of ironies, and not the least is that the transformation of the British economy owed much to the work of men whose place in British society was less than exalted. Very few of the British engineers listed in this and the previous chapter came from the upper classes. Only one, Thomas Savery, came from a landowning family, while a few, most notably Nasmyth, were the products of families of some professional standing. In contrast, the lower end of the social scale is quite well represented. It is also noteworthy that many of the men who made decisive contributions to Britain's rise as a great industrial power were themselves products of the countryside. In the elegant phrasing of L.T.C. Rolt, they[56]

> ...were drawn to the new workshops from remote rural areas, not by the pressure of necessity, but by a mysterious aptitude for and love of mechanics. But is this, after all, any stranger than the inexplicable flowering of poetic or artistic genius? For the truth is, surely, that these pioneer engineers were the artists of their profession whose careers were determined by the artist's compulsive need to fulfill his creative endowment.

Some engineers were able to parlay their technical skills into the ownership of substantial enterprises. In some cases, as with Watt, they were greatly aided in this endeavor by their association with already successful businessmen. Others were able to start their business largely through their own efforts. For the mechanical engineers that were beginning to emerge in the 19th century, this endeavor was facilitated by the possibility of starting small. Maudsley, Whitworth, and Fairbairn all began with tiny workshops and one or two assistants.[57] Roberts, at least according to one story, started with single lathe that was driven from the basement by his wife.[58] Other, less renowned engineers were similarly able to establish small enterprises; by the middle of the 19th century, 677 machine-building firms could be found in Britain, of which two-thirds had fewer than ten employees, while only 14 had more than 350.[59] In Francois Crouzet's summation, "engineering had many small masters... and a limited number of 'true' industrialists."[60]

Recognition of engineers' accomplishments in the form of a handful of knighthoods and baronetcies began to come only in the middle years of the 19th century, when the industrialization of Britain was well established, and Britain was in fact about to decline relative to newly industrializing nations.

Living and working in the first industrial nation gave British engineers many favorable opportunities, even though their government did little to promote engineering. By contrast, the French government actively encouraged technological progress, but its efforts may have had the ironic result of retarding the development of French engineering. Despite a roster of scientists and engineers that was unmatched anywhere in the 19th century, France lagged behind her cross-channel neighbor in the scale and pace of industrialization. Although the reasons for this differential are complex, part of the answer may lie in a government-supported system of engineering education that was strongly inclined towards mathematical and theoretical approaches, with a consequent devaluation of the empirically based engineering at which British engineers were so successful. Moreover, the centralized nature of the French state resulted in a much greater degree of political, economic, and cultural concentration than was to be found in Britain. While engineering talent could find outlets throughout Britain, in France technical proficiency had a strong tendency to gravitate to Paris. The French capital, like all the capital cities of Europe, was not an important industrial center, so this siphoning off of engineering talent may have deprived other regions of the technical skills they needed, while at the same time denying French engineers the on-the-job training that actual involvement with industry could provide.[61]

British engineers made their country "the workshop of the world" for a century, helped to create a modern industrial society, and in so doing transformed the whole of civilization. From their own country British engineers spread industrial technology across the globe. Yet their reign did not go unchallenged. As the 19th century wore on it was becoming evident that other nations were producing engineers of equal and sometimes greater talent. One of these nations had only recently gained its independence from colonial rule and was laying the foundations of an industrial economy in a largely agrarian frontier country. To the surprise of many Europeans, the United States of America was beginning to assert itself as an economic and technological power. Many of its successes were the result of the efforts of an emerging group of engineers. It is to their story that we will now turn.

Notes

1 Jeffrey G. Williamson, "Why Was British Growth So Slow during the Industrial Revolution?" *Journal of Economic History* 44, 3 (September 1984), pp. 688-89. The growth rate for the economy as a whole was of course influenced by the growth of the industrial sector, but several decades were to pass before industry comprised a substantial part of the total economy.

2 For an anthology of recent thinking on the causes of the industrial revolution, see Joel Mokyr, "The Industrial Revolution and the New Economic History," in Joel Mokyr (ed.), *The Economics of the Industrial Revolution* (Totowa, New Jersey: Rowman and Allenheld, 1985)

3 M.W. Flinn, *The Origins of the Industrial Revolution* (London: Longmans, Green, and Co., 1966), p. 102.

4 R.M. Hartwell, *The Industrial Revolution and Economic Growth* (London: Methuen, 1971), pp. 286-87.

5 Mokyr, op. cit., pp. 28-29.

6 Much as we would like to avoid it, the use of the third person masculine is appropriate, for only men were mechanical engineers during this period. We will take up the subject of women engineers in chapter 11.

7 Samuel Smiles, *The Lives of the Engineers* (London: John Murray, 1904), p. xxiii.

8 A.E. Musson and E. Robinson, "Science and Industry in the Late Eighteenth Century," *Economic History Review* 2nd series, 13 (1960), p. 429.

9 L.T.C. Rolt, *Thomas Newcomen: The Prehistory of the Steam Engine* (Dawlish, Devon: David and Charles, 1963), pp. 35-41.

10 Aubrey Burstall, *A History of Mechanical Engineering* (Cambridge, Massachusetts: MIT Press, 1965), p. 193.

11 Rolt, op. cit., p. 4.

12 Ibid., p. 48.

13 J.R. Harris, "The Employment of Steam Power in the Eighteenth Century," *History* 52, 175 (1967), p. 147.

14 Rolt, op. cit., p. 73.

15 Useful studies of Watt and his work are D.S.L. Cardwell, *Steam Power in the Eighteenth Century* (London: Sheed and Ward, 1963), Henry W. Dickenson, *James Watt: Craftsman and Engineer* (New York: Augustus M. Kelley, 1968) and L.T.C. Rolt, *James Watt* (London: B.T. Batsford, 1962).

16 Paul Mantoux, *The Industrial Revolution in the Eighteenth Century* (New York: Harper and Row, 1962), p. 323.

17 A similar device had been invented four years earlier at the Woolwich Arsenal by Jan Verbruggen. The fact that Wilkenson received a patent once again demonstrates the capriciousness of the British patent system.

18 Abbott Payson Usher, *A History of Mechanical Inventions* (Cambridge, Massachusetts: Harvard University Press, 1954), p. 359

19 David S. Landes, *The Unbound Prometheus: Technological Changes 1750 to Present* (Cambridge: At the University Press, 1969), p. 103.

20 D.S.L. Cardwell, *Turning Points in Western Technology* (New York: Science History Publications, 1972), pp. 88-89.

21 Peter Mathias, Who Unbound Prometheus? Science and Technical Change, 1600-1800," in Peter Mathias, *Science and Society, 1600-1900* (Cambridge: At the Univeristy Press, 1972), p. 138.

22 Donald Fleming, "Latent Heat and the Invention of the Watt Engine," in Otto Mayr (ed.), *Philosophers and Machines* (New York: Science History Publications, 1972), p. 123.

23 Robert E. Schofield, *The Lunar Society of Birmingham: A Social History of Provincial Science and Industry in Eighteenth Century England* (Oxford: At the Clarendon Press, 1963), pp. 145-46.

24 Ibid., p. 438.

25 Musson and Robinson, op. cit., pp. 243-44.

26 Rolt, *James Watt*, op. cit., pp. 72-73 and 122-24.

27 Rolt, *Thomas Newcomen*, op. cit., p. 79.

28 Mokyr, op. cit., p. 6.

29 Cardwell, *Turning Points*, op. cit., p. 116.

30 Usher, op. cit., p. 359.

31 L.T.C. Rolt, *A Short History of Machine Tools* (Cambridge, Massachusetts, MIT Press, 1965), p. 53.

32 Francois Crouzet, *The First Industrialists: The Problem of Origins* (Cambridge: Cambridge University Press, 1985), pp. 91-92.

33 Usher, op. cit., p. 367.

34 Robert S. Woodbury, *History of the Lathe to 1850* (Cleveland: Society for the History of Technology, 1961), p. 14.

35 Cardwell, *Turning Points*, op. cit., p. 117.

36 Burstall, op. cit., pp. 225.

37 Usher, op. cit., p. 378.

38 For Marc Brunel's early life, see L.T.C. Rolt, *Isambard Kingdom Brunel: A Biography* (London: Longmans, 1957) and Paul Clements, Marc I. Brunel (London: Longmans, 1970).

39 Sigvard Strandh, *The History of the Machine* (New York: Dorset Press, 1989), p. 56.

40 John B. Rae, "The Rationalization of Production," in Melvin Kranzberg and Carroll S. Pursell (eds.), *Technology in Western Civilization*, vol. 2 (London: Oxford University Press, 1967), p. 40.

41 E.A. Forward, "Samuel Goodrich," Newcomen Society *Transactions*, 3 (1922-23), pp. 7-8.

42 Rolt, *Isambard Kingdom Brunel*, op. cit., p. 22.

43 L.T.C. Rolt, *Victorian Engineering* (Harmondsworth: Penguin, 1970), p. 242.

44 *Dictionary of National Biography*, ed. Leslie Stephen and Sidney Lee (Oxford: at the University Press, 1917) vol. 16, pp. 1276-77.

45 Woodbury, op. cit., pp. 108-112.

46 L.T.C. Rolt, *A Short History of Machine Tools* (Cambridge, Massachusetts, MIT Press, 1965), p. 106.

47 Ibid., p. 107.

48 James Nasmyth, *Engineer: An Autobiography* ed. Samuel Smiles (London: John Murray, 1905), p. 110.

49 Ibid., p. 89.

50 Ibid., pp. 231.

51 Nathan Rosenberg, *Perspectives on Technology* (Armonk, New York and London: M.E. Sharpe, 1985), p. 120.

52 *Dictionary of National Biography*, op. cit., vol. 21, pp. 166-69.

53 J.D. Scott, *Vickers: A History* (London: Weidenfeld and Nicholson, 1962), p. 27.

54 *Dictionary of National Biography*, op. cit., vol. 6, pp. 987-88.

55 J.W. Roe, *English and American Tool Builders* (New York: McGraw-Hill, 1926), p. 72.

56 Rolt, *A Short History of Machine Tools*, op. cit., pp. 105-106.

57 Crouzet, op. cit., p. 90.

58 Rolt, *A Short History of Machine Tools*, op. cit., p. 106.

59 Crouzet, op. cit., p. 35.

60 Ibid.

61 Cardwell, *Turning Points*, op. cit., pp. 122-27.

Chapter 7

The Founding of American Engineering

During the last quarter of the eighteenth century, Britain was losing a major part of its first colonial empire just as the advancing pace of industrialization was making it the world's greatest economic power. The loss of the American colonies did not create much in the way of economic damage, and Britain went along its way toward becoming the workshop of the world. On both sides of the Atlantic few thought that in due time the former colonies would challenge Britain for industrial leadership. The newly formed United States was a predominantly agrarian nation, thinly populated, and lacking many of the trappings of culture found in more civilized realms. But in the ensuing decades the new country gained in wealth and power as rivers were spanned, towns were connected by canals, and abundant resources began to be tapped. At the same time, American industrial products began to gain worldwide respect for their durability and low price. All of the these endeavors required engineering efforts that often verged on the heroic. The story of nineteenth-century America's rise to economic and political prominence is also the story of the emergence of its engineers.

American engineers of this era shared with their contemporaries in Europe the advantage of living at a time when vital statistics and biographical data were kept with reasonable accuracy. In contrast to engineers of earlier times, we know who they were as well as what they did. While there were many connections between American and European engineers and engineering, the American situation was just different enough to warrant separate consideration. The uniqueness of the American experience also demonstrates how engineering can be shaped by the special requirements of time and place.

Although much of America was still a primitive, frontier territory, American engineers generally enjoyed a favorable environment. They worked in a new and rapidly expanding country where there were less rigid social divisions than in the old country. There was a vast amount of engineering work that needed to be done, and those who could do it were valued members of their society. When the United States emerged as a nation the number of trained engineers was minuscule, and for at least the next century the supply of such engineers remained well short of the demand. This held true even when "trained" is used in a broad sense to include not only academically educated engineers but also those who had some sort of systematic training in the fundamentals of engineering through apprenticeship or even through self-education.

The wide variety of men who set themselves to engineering tasks produced an unfortunate confusion in the American perception of what an engineer really was. There grew a strong tradition of "Yankee ingenuity" borne by men who had little formal training and even less respect for it. Accordingly, the heroes of American technology include individuals like Fulton, McCormick, and later Edison and Ford—self-taught inventors who often took pride in their lack of book learning. Second, the term "engineer" came to be applied to occupations that had scant relation to professional engineering. Conspicuously, in the United States the operator of a locomotive is an engineer, whereas in Great Britain he is more properly known as an engine-driver. There have been other, even more outlandish uses of the word, as when a garbage collector becomes a "sanitary engineer."

None of these linguistic aberrations did much to retard the development of American technology. Their effect was to give engineers less credit than they deserved for technological advance, and to obscure engineering's place as a distinct occupation with its own methods and standards. At the same time, there always was the recognition that formally trained engineers were a national asset. At the beginning of the nineteenth century it was not clear just what constituted an appropriate degree of training, but there was little doubt about its desirability. Still, ambiguities remained, and American engineering drew from both the British tradition of on-the-job training and the French system of formal academic education.

Early American Engineers

In the period immediately following the Revolutionary War, the fledgling United States remained heavily dependent on European engineers for the design and construction of works that were beyond the capacity of skilled native craftsmen. A reliable estimate puts the number of people that could be considered qualified engineers in 1816 at not over thirty.[1] As a rule the Europeans did some engineering work and then returned home, as with Marc Brunel, whom we met in the last chapter. Another was William Weston (1752-1853), an Englishman who may have worked with James Brindley. After arriving in the United States in 1793, he was in considerable demand as a consultant on canal projects, among them a project organized by George Washington to construct a set of locks at Great Falls on the Potomac. At the time, American canal builders were ignorant of such basics as how to accurately survey vertical distances, seal canal beds, and construct workable locks.[2] Weston's advice and supervision, brief though it was in most cases, was essential to American civil engineering. He returned to England in 1800 and a few years later refused an offer to become chief engineer of the Erie Canal. A third was Lewis Wernwag (1769-1843), a German who came to America in 1786 and engaged in a variety of engineering operations.[3] With a versatility characteristic of many 18th century engineers, he built bridges, worked on developing anthracite as a fuel, and constructed mills and machines. He ended up an American engineer, for unlike Brunel and Weston he stayed in the United States.

In spite of their limited numbers, American-born engineers compared favorably with European ones. One of the best of them was Laommi Baldwin (1745-1807). The son of a carpenter in Woburn, Massachusetts, he received a common school education and was apprenticed to a cabinet maker.[4] Not content with the training he was receiving, he attended weekly lectures at Harvard on mathematics and physics that were given by Professor John Winthrop. To get to these classes, Baldwin walked from Woburn to Cambridge and back (about ten miles each way) accompanied by his fellow townsman, Benjamin Thompson, the future Count Rumford.

At the age of twenty Baldwin became a surveyor, a common method of entering engineering in those days. His career was briefly interrupted by service in the Revolutionary War, but he went on to become one of the first native-born Americans who could claim the title civil engineer. His most substantial achievement was the construction of the Middlesex Canal, built with private funds (and the essential assistance of William Weston) to connect Boston with the Merrimac River at approximately the present site of the city of Lowell.

Baldwin's son, Laommi Baldwin, Jr. (1780-1838) graduated from Harvard in 1800. He studied law and practiced until 1807, but turned to civil engineering, where he was primarily employed in the construction of canals and harbors.[5] For our purposes, he is particularly interesting for his views on the proper status of an engineer. He resigned from his position as chief engineer of the Union Canal in Pennsylvania when the company's directors overruled him on his choice of the canal's route and dimensions. In explaining his resignation, he rejected the notion that "the engineer, as engineer, could be a mere technician"[6] whose sole responsibility was to execute the decisions of the board of directors. The engineer, he believed, should make the contracts. He should have a "good and judicious plan of operations" and above all be a person of "wisdom, talent, and disinterested devotion to the public good."

The issue that Baldwin raised about the professional status of the engineer has still to be resolved, and it will be one of the chief issues taken up in Chapter 11. We can find a parallel between Baldwin's attitude and that of I.K. Brunel, and for much the same reason. Both were strong-minded individuals, with a strong sense of their superiority. Baldwin took this to an absurd degree when he remarked in 1826 that "As far as I can see and learn, there are as yet no civil engineers in the United States of America"[7] — presumably excepting himself.

Benjamin Banaker (1731-1806) was more an astronomer and mathematician than an engineer, but he is listed in the *Dictionary of American Civil Engineers,* so he merits at least a few sentences here. His background makes him unique among early engineers, for he was the son of a freed slave and a woman of mixed race.[8] His grandmother, an Englishwoman who had arrived in America as an indentured servant, gave him his early education and he attended a school run by a local Quaker farmer. While he worked his father's tobacco farm he taught himself surveying. The farm was located near the site outside Baltimore where John and Andrew Ellicott began to build mechanized grist mills in 1771. Banaker's keen interest in the mills eventually led to his becoming an assistant to Andrew Ellicott when the latter went on to lay out the new district of Columbia in 1791, thereby implementing Pierre L'Enfant's plan for the national capital.

If the elder Laommi Baldwin can be accepted as the first identifiable American civil engineer, then Oliver Evans (1755-1819) certainly holds the same place among American mechanical engineers. He was born in Newport, Delaware, the son of a farmer who had earlier been a shoemaker. Apprenticed to a wheelwright at 16, he eventually became a reasonably successful storekeeper in Newport.[9] Towards the end of the Revolutionary War he conceived the idea of a completely mechanized grist mill, which was eventually completed in 1785. Powered by water wheels, through a combination of hoists, conveyors, chutes and other devices it converted the grain into flour without the intervention of human labor.

When he became interested in the potentialities of steam power he moved to Philadelphia, where he felt that opportunities were better than in largely rural Delaware. There he engaged in various kinds of manufacturing projects, with steam engines being the most important. He did a good deal of experimenting with steam at higher than atmospheric pressures, but he never realized his dream of operating "steam waggons" on American roads, except for the brief jaunt of his steam-powered dredger through the streets of Philadelphia.

He could undoubtedly have built the vehicles, but at that time there were very few roads in the United States that they could have run on.

Evans' background and training were very similar to his British 18th century contemporaries. Wheelwrights, like millwrights and masons, were one of the principal sources of engineering talent in both Europe and America, and Evans, like the most distinguished of his British counterparts, did a substantial amount of self-education. He had fewer opportunities to utilize his genius, because in his lifetime the United States was experiencing only the first murmurs of the Industrial Revolution; it can truly be said that he was a man born too early.

Evans was unquestionably a genius, but it is difficult to access his actual impact on American engineering. His grist mill established a pattern that was generally followed in the American flour milling industry. It can even be seen as the prototype for mechanized process industry in general. His work with steam engines showed boldness and initiative; had he lived just a few years later, when the railroad appeared on the American scene he might well have become the American George Stephenson. Here would have been the place to use the "steam waggons" that he dreamed of and tried so vigorously to promote.

Civil Engineers

For at least the first half of the nineteenth century the greatest demand for engineers in the United States was in the civil field, and understandably so. The country was expanding rapidly and clamoring for "internal improvements"—roads, canals, harbors, and most of all railroads. Like the British civil engineers under comparable conditions, the Americans worked in all existing fields of engineering. The demand for engineers far outran the supply, a situation that made it attractive for ambitious young men to enter the occupation.

These engineers represented a cross-section of American society and entered the field through a variety of channels. Until mid-century, virtually the only academic route was the

United States Military Academy at West Point, which up to then had produced about fifteen percent of the nation's engineers.[10] The Corps of Engineers was the elite branch of the Army, attracting the Academy's top graduates, notably Robert E. Lee and George McClellan. Army engineers were routinely assigned to internal improvement projects, and it was common practice for a young man to secure an appointment at West Point with the intention of going into private engineering practice as soon as he discharged his Army obligation. This was understood and approved. During the Age of Jackson there was an egalitarian desire to avoid the formation of a permanent officer caste produced by military academies; officers were expected to rise from the ranks. Nevertheless, the Army Military Academy at West Point was successfully defended on the ground that it was an indispensable source of needed non-military engineers.[11]

There would be little point in attempting to enumerate all the West Pointers who contributed to American civil engineering during this period; a few examples will suffice. Stephen H. Long (1784-1864) was one of the most distinguished of those who stayed in the Army but engaged extensively in civil works.[12] Long attended Dartmouth College, joined the Army as an engineer officer and for a time taught mathematics at West Point. He then engaged in extensive exploration and surveying in the western parts of the country, getting as far as the Rocky Mountains. He was later, in accordance with the practice of the time, assigned to assist with the construction of the Baltimore and Ohio Railroad, and located the route in cooperation with a self-taught engineer named Jonathan Knight (1787-1858). He also surveyed the route of the Chesapeake and Delaware Canal, and eventually rose to being Chief of the Topographical Engineers Corps.

Of the West Point graduates who had distinguished engineering careers in civilian life during the first half of the nineteenth century, two can be selected as especially interesting illustrations because of their close professional and personal relationships. They are William Gibbs McNeill and George Washington Whistler.[13] Both were the sons of British army officers who had returned to America after the Revolutionary

War. McNeill's father became a physician, Whistler's an officer in the U.S. Army. Both were educated at West Point, McNeill arriving two years before Whistler. Upon graduation they were assigned to internal improvement projects. McNeill helped to survey canals and then went to work on railroads. He and Whistler were sent to Britain to study railroad construction and management, where they got to know Telford and Stephenson.

McNeill resigned from the army in 1837 to devote himself to railroad engineering. His masonry viaduct in Massachusetts is still in service on the mainline of what today is known as the Northeast Corridor. On a visit to Europe in 1851 two years before his death he became the first American elected to the British Institution of Civil Engineers.[14]

Whistler left the army in 1833 after an active career in railroad building to become superintendent of the Locks and Canals Company in Lowell, Massachusetts. In the company's machine shop he built locomotives, demonstrating the close connection then to be found between civil and mechanical engineering. He subsequently went to Russia to build that country's first major railroad, the line between St. Petersburg and Moscow. He went on to do more engineering work in Russia, where he died in a cholera epidemic. The railroad was finished by his son George.

Whistler and McNeill had another relationship, one that eventually immortalized the name of the former. Whistler was married to McNeill's sister, and their first child went on to a distinguished career as a painter. His most famous portrait, "Arrangement in Gray and Black," is better known by its popular title, "Whistler's Mother."

Although West Point was the sole source of formally trained engineers, many capable engineers learned their trade through the time-honored method of learning on the job. One of the most accomplished products of this pattern was Canvass White, one of the builders of the first great American engineering project, the Erie Canal. DeWitt Clinton is often given credit for the construction of the Erie Canal, and he deserves his place as the man who promoted it and persuaded the state of New York to undertake it, but the actual building was directed by White and two others, James Geddes and Benjamin Wright. All were men of substance, and all had done extensive surveying. On the strength of their prior experience they were given the monumental task of laying out and overseeing the construction of a 300-mile ditch from the Hudson River to Lake Erie. If they were not bonafide engineers when they began, they certainly were by the time that they had finished. All three went on to further achievements in civil engineering after they completed the Erie Canal. Among his other projects, Canvass White took over the construction of the Union Canal after Laommi Baldwin's angry resignation.

The magnitude of the Erie Canal enterprise required more engineers than were available in the United States at the time. The only recourse was extensive on-the-job training, so the canal became a school for engineers at least equal in importance to West Point. Young men were employed to assist surveying crews or to help cut through terrain that was still largely wilderness, and if they had sufficient ambition and talent, they emerged as full-fledged engineers.

The outstanding example of this process was John B. Jervis (1795-1885), frequently regarded as the greatest of mid-nineteenth century American civil engineers.[15] A carpenter's son from Huntington, New York, he was employed to clear brush and cut timber for the surveyors of the Erie Canal. He rose rapidly to become a member of the survey crew and by 1819 he had been appointed resident engineer for a section of the canal. From the Erie he went on to be chief engineer of the Delaware and Hudson Canal (the point where it reached the Delaware River became Port Jervis, N.Y.).

Jervis subsequently moved into railroad building, where he demonstrated outstanding skill as a mechanical engineer. One of his most important inventions illustrates the ability of a successful engineer to respond to specific circumstances. The construction of early American railroads was usually a slapdash affair; builders were eager to traverse the wild territory of the new nation, and they lacked the time and money to construct a substantial right of way. The result was sharp curves, steep grades, and uneven trackwork. Under these circumstances, locomotives of British design bounced, derailed, and rapidly wore out the track. Jervis' solution was elegantly simple. He fitted a locomotive with swiveling four-wheel lead truck, which by guiding the locomotive into a curve markedly increased its ability to negotiate the tortuous trackwork characteristic of American railroads.[16]

After years of working on canals and railroads, Jervis turned to hydraulic engineering, serving as chief engineer for the Croton Aqueduct, which provided the water supply for New York City. Jervis' career shows that in his lifetime American engineering was still not specialized to any great degree. At the same time, the way he went about his work provides one of the best examples of the direction in which American engineering was evolving. Although his engineering was almost wholly learned on the job, Jervis was firmly committed to systematizing and regularizing the practice of engineering, whether through calculating the tractive force of a locomotive, finding a formula for redesigning a canal prism, or developing a new mathematical constant to aid in the construction of reservoirs.[17] As exemplified by Jervis, engineers were beginning to move away from long-standing procedures that did not always result in good engineering practice.

Charles Ellet, Jr. (1810-1862) offers a variation on this pattern. His father had been a hardware merchant in Philadelphia, but just before Charles was born he had become a farmer. Charles had little formal schooling, but his talented mother encouraged his aptitude for mathematics.[18] In 1827 he went to work for a surveying crew on the North Branch of the Susquehanna River, with the title of Assistant Engineer. The project was part of Pennsylvania's ambitious Main Line project of canals and railroads that the state hoped would compensate for the advantages the Erie Canal was giving New York. Two of the Erie Canal's main builders, James Geddes and Canvass White, were also associated with this work, and they set up the same system for training engineers as they had used on the Erie.

After a year on the Pennsylvania waterway, Ellet spent two years working under Benjamin Wright (the third member of the Erie Canal engineering triumverate) on the Chesapeake and Ohio Canal. He then decided that his professional skills would be improved by more formal training, and accordingly he spent a year at the Ecole des Ponts et Chaussees in France. When he returned he embarked upon a career that included being the chief engineer of the James River and Kanawha Canal, and building two suspension bridges, one over the Schuylkill River at Philadelphia and the other over the Ohio at Wheeling.

The feat for which he is best known is building steam-power rams for use in naval warfare. He had tried to persuade the Russian government to adopt this weapon during the Crimean War, and when the Civil War came along he badgered the Union government into building such vessels for use on the Mississippi. Two of them, with Ellet in command, took part in the battle for Memphis on 6 June 1862, and one did succeed in ramming and sinking a Confederate warship. The value of the rams in the winning of the battle remains controversial, but this didn't matter to Ellet, for ten days later he died of a wound suffered in the battle.

Another American engineer who chose to study in Europe was Moncure Robinson (1802-1891). The son of a merchant in Richmond, Virginia, He attended William and Mary College for three years (beginning at the age of thirteen) but left without a degree, and volunteered in 1819 to work on surveying the route of the James River and Kanawha Canal.[19] In 1822 he went on a journey to examine the Erie Canal, and a year later was appointed engineer for the State of Virginia. In 1825 he left for Europe and studied at the Sorbonne for the winter sessions of 1825-26 and 1826-27. Having witnessed the beginning of railroad construction in Britain, he returned home full of enthusiasm for the new mode of transportation. He helped to build the Richmond, Fredericksburg, and Potomac Railroad, and eventually became its president, one of the earliest in a line of engineers who moved into railroad management.[20]

Native-born American engineers continued to be supplemented by immigrants. One of the greatest was Benjamin

Henry Latrobe (1764-1820). The English-born son of a Moravian minister, he was sent to Germany for his education. He attended a Moravian seminary and then the University of Leipzig, and even served for a time in the Prussian cavalry.[21] When he returned to England he studied both architecture and engineering, the latter under John Smeaton. He came to the United States in 1796. His career exemplified the long-standing tradition of the association of architecture and civil engineering. His best-known works were architectural, but he was also active in canal building, steamboat construction on the Ohio River, and water supply systems.[22] Sadly, he died of yellow fever after he had gone to New Orleans to finish a water supply system that had been begun by his son Henry, who had also died of Yellow fever. Another son, Benjamin Henry Latrobe, Jr. (1806-1878) practiced law for a time (like Laommi Baldwin, Jr.) before turning to engineering. He achieved some distinction in building railroads. He became chief engineer to the Baltimore and Ohio Railroad and originated the ton-mile as the unit of work in railroad operation.

The lives and accomplishments of native and immigrant engineers could be continued at length if there were any purpose served by doing so. The sheer volume of engineering work to be done in the building of what we would today call infrastructure was bound to draw large number into civil engineering, and for the 19th century the personal data are available to identify them with a precision that is impossible for earlier eras. The accompanying tables, which are based on the Archive of American Engineers in the Smithsonian Institution's National Museum of American History, give some overview of the characteristics of American engineers born before 1860. The figures are incomplete, but they give an indication of how the proportion of engineers from farm and craftsman families declined, while the proportion from business and professional families increased correspondingly. The second table also shows a striking increase of academically trained engineers among the group born after 1831. Most of these would have completed their education when facilities for the higher education of engineers were proliferating rapidly.

Table 1
The Social and Economic Backgrounds of 19th Century American Engineers

	Dates of Birth		
	before 1790	1790-1830	1831-60
Family background: (percentages)[23]			
Farm	31.3	17.4	13.7
Business	19.6	33.0	31.0
Professional (law, medicine, clergy)	9.8	21.4	30.4
Craftsman	18.2	11.0	6.0
Military	13.7	7.0	5.5
Engineer	3.9	6.3	9.0
Types of training			
Self-taught	17.4	12.0	3.0
Apprenticeship	22.8	18.0	14.7
Non-Technical Higher Education			
U.S.	8.7	10.4	9.0
Other	2.2	–	.4
Higher Education in Engineering and Science			
U.S.	3.3	12.8	44.0
Other	1.1	3.3	6.8

Irrespective of their social status or educational experiences, American engineers created a distinct style of engineering. This style was shaped by the particular circumstances in which they worked; America was not Europe, and the standards of good engineering reflected the differences between the New World and the Old. The interplay between native conditions and the way that American engineers went about their work was clearly delineated by David Stevenson, a British engineer who had traveled extensively in the United States:[24]

> At the first view, one is struck by the temporary and apparently unfinished state of many of the American works, and is very apt, before inquiring into the subject, to impute to want of ability what turns out, on investigation, to be a judicious and ingenious arrangement to suit the circumstances of a new country, of which the climate is severe—a country where stone is scarce and wood is plentiful, and where manual labor is very expensive. It is vain to look to the American works for the finish that characterizes those of France, or the stability for which those of Britain are famed. Undressed slopes of cuttings and embankments, roughly built rubble arches, stone parapet-walls coped with timber, and canal locks wholly constructed of that material, every where offend the

eye accustomed to European workmanship. But it must not be supposed that this arises from want of knowledge of the principles of engineering, or of the skill to do them justice in the execution. The use of wood, for example, which may be considered by many as wholly inapplicable to the construction of canal-locks, where it must not only encounter the tear and wear occasioned by lockage of vessels, but must be subject to the destructive consequences of alternate immersion in water and exposure to the atmosphere, is yet the result of deliberate judgment. The Americans have, in many cases, been induced to use the material of the country, ill adapted though it may be in some respects to the purposes to which it is applied, in order to meet the wants of a rising community, by speedily and perhaps superficially completing a work of importance, which would be otherwise delayed, from a want of the means to execute it in a more substantial manner; and although the works are wanting in finish, and even in solidity, they do not fail for many years to serve the purposes for which they were constructed, as efficiently as works of a more lasting description.

If Stevenson had visited the United States a generation later one wonders what he would have thought of Herman Haupt's Civil War railroad trestles, described admiringly by President Lincoln as looking as though they had been made of "nothing but bean poles and cornstalks." Haupt (1817-1905) was a West Pointer with a long career in railroad construction and management. He too demonstrates close connection between civil and mechanical engineering to be found in this era; for the building of the Hoosac Tunnel in Massachusetts he designed a pneumatic drill superior to any other then in use. During the Civil War he supervised railroad operations for the Army of the Potomac, and later, foreshadowing the direction of engineering education, he was for a time Professor of Civil Engineering at the University of Pennsylvania.[25]

Mechanical Engineers

As we have just noted, for much of the 19th century there was little distinction between civil and mechanical engineers. The American Society of Mechanical Engineers was not founded until 1880, and even afterwards many civil engineers continued to be all-purpose engineers. Jervis and Whistler built locomotives as well as railroad roadbeds; Latrobe the younger built steam boats; Ellet designed rams; and James B. Eads (1820-1887), who built the steel arch bridge that still carries traffic over the Mississippi at St. Louis, also constructed ironclads for use in the Civil War, in which for the first time heavy guns were trained by steam power.[26]

At the same time however, there was also a growing body of American engineers who could be more accurately classified as mechanical rather than civil engineers. One of the most prominent was George H. Corliss (1817-1888). A country doctor's son who early showed mechanical and mathematical talent, Corliss attended Castleton Academy in Vermont for three years and then became a storekeeper.[27] Because of customer complaints about poorly made footwear he experimented with a sewing machine for attaching soles to uppers, but gave it up to work in a machine shop in Providence, Rhode Island. He established the Corliss Machine Shop in Providence in 1848 and made himself the greatest of the nineteenth century American steam engineers. He developed a system of rotary valves for controlling the admission of steam to the cylinders so that the engines ran at a constant speed. His machine shop built the power plant for the U.S.S. *Monitor*, and in 1876 he was asked to provide power and heat for the Philadelphia Centennial Exhibition. He did so with a single giant machine that weighed 700 tons and produced 1400 horsepower. This engine remained his property, and in spite of pleas that Sunday was the only day many people could attend the Exposition, he refused to allow it to be run on the Sabbath.

Despite the steady advance of native-born engineers, immigrant engineers continued to play an important role. One British example is James Bicheno Francis (1815-1892), who could also be classed as a civil engineer, and in fact served as president of the American Society of Civil Engineers from 1880 to 1882. However, he is properly placed among the mechanicals because the bulk of his career was devoted to hydraulics and the use of water power.[28] His father was superintendent of a port, canal, and railroad company in South Wales, and James went to work for him by the time he was fourteen. He emigrated when he was eighteen, worked on railroad construction with Whistler and McNeill, and then went to Lowell, Massachusetts to work with Whistler for the Proprietors of Locks and Canals.

The Proprietors of Locks and Canals was a company formed in the 1790s to improve navigation on the Merrimac River. When Lowell became a center of textile manufacturing in the 1820s, the company passed into the control of the mill owners, who used it for allocating and regulating water power. Francis became a friend of Uriah Boyden (1804-1879), who had invented an improved water turbine for use in the Lowell mills. Francis went well beyond his friend's design; in 1851 he developed a markedly successful inward-flow turbine. Its basic design, which bears his name, is still employed today. Equally important, Francis committed himself to the systematic study of hydraulics. Although he had very little formal schooling, he approached engineering with a thoroughly scientific spirit, and always sought to communicate his discoveries to other engineers through his publication of over 200 technical papers, most of them on the flow of water. The formula for determining the flow of water over a weir is named after him, while the procedures used for designing turbines continue to be based on his principles.[29]

Several of the engineers previously listed have been mentioned as designers and builders of locomotives. To them must be added Mathias Baldwin (1795-1866), founder of what became the county's largest manufacturer of steam locomotives, and one of the largest of the nineteenth-century industrial enterprises. His father was a carriage maker in Philadelphia. After his death his property was lost by the executor when Matthias was four. Matthias was eventually apprenticed to a jeweler, but he disliked the jewelry business and turned to making tools for engineers and printers. In 1831 he took up locomotive building; his first product, constructed with tools of his own making, was an engine named *Old Ironsides*. He found a ready market, for most early American railroads were too small to build their own locomotives, as was customary in Europe, and were quite willing to buy locomotives of standard design "off the shelf." Baldwin had to take into account the same conditions that had motivated Jervis' key invention; among Baldwin's important inventions was the flexible-beam driver truck, which allowed engines with long wheelbases to negotiate curves more easily.[30]

A crucial source of American engineering progress was the development of the machine tool industry, which was surprisingly advanced and extensive for a country that until after the Civil War was behind Europe in its general technological development. The development of a machine tool industry can be attributed to the growth of the "American system of manufacturing," as it was designated by European industrialists by 1850. With its emphasis on standardization and interchangeability, the American system created a strong and apparently endless demand for specialized tools.[31]

Just as canal and railroad building provided schools for civil engineers, so the machine shops became a training ground for engineers. One of the most noteworthy examples is William Sellers (1824-1905). A member of a well-to-do family, he attended private school before going through an apprenticeship as a machinist. He subsequently founded a machine shop in Philadelphia which was devoted to the production of machine tools.[32] In one important way he was the American version of Joseph Whitworth, for his greatest achievement was to propose a standard screw thread that was adopted by the United States and eventually throughout the world. The thread that Sellers designed was superior to Whitworth's in that it could be more easily duplicated by standard shop practices.

In this endeavor, Sellers had the strong organizational support of Philadelphia's Franklin Institute, of which he served as president. A special committee of the Institute actively promoted the new set of screw threads through an appeal to the federal government and private enterprises. It was successful because Sellers had in fact designed a superior thread, but equally important, the Institute enjoyed the respect and support of the American business community. The committee was staffed by prominent members of the Philadelphia's ironworking industries, while the increasing interdependence of America's industrial establishments made a uniform set of standards increasingly necessary.[33] Sellers' successful design of a new screw thread underscored the growing importance of organizations — professional societies like the Franklin Institute and large industrial enterprises—in defining the work of the engineer.

In 1874 Sellers took charge of a steel company that was renamed the Midvale Steel Company. This was to have important consequences for the changing organizational context of American engineering, for it was here that a young engineer named Frederick Winslow Taylor began a series of experiments in cutting metals that had a significant impact on machining practices. These were not the full extent of Taylor's influence; as we shall see in Chapter 9, Taylor had strong ideas about the role of engineers in modern organizations, ideas that were to leave their mark on both the management of organizations and the practice of engineering.

American engineers had played an indispensable role in transferring the products and processes of the British Industrial Revolution to their own country, adapting them to local conditions, and frequently improving them in the process. But as the American economy and society began to change dramatically, a new set of challenges appeared. During the last quarter of the nineteenth century engineering in the United States and elsewhere was undergoing sweeping changes in terms of educational standards, specialization, professionalization, and the kind of duties taken on by engineers. The nature of these changes and their connection to the practice of engineering will be the topic of the chapters to follow.

Notes

1. Daniel C. Calhoun, *The American Civil Engineer* (Cambridge, Massachusetts: MIT Press, 1960), p. 22.

2. Elting W. Morison, *From Know-How to Nowhere: The Development of American Technology* (New York: New American Library, 1977), pp. 22-25.

3. *A Biographical Dictionary of American Civil Engineers* (New York: American Society of Civil Engineers, 1972), p. 124.

4. *Dictionary of American Biography*, Allen Johnson, ed. (New York: Charles Scribner's Sons, 1928-58), vol. 1, pp. 539-540.

5. Calhoun, op. cit., p. 98.

6. Ibid.

7. Ibid., p. 263.

8. Silvio A. Bendini, *The Life of Benjamin Banneker* (New York: Charles Scribner's Sons, 1972)

9. Greville and Dorothy Bathe, *Oliver Evans: A Chronicle of Early American Engineering* (Pennsylvania: Historical Society of Pennsylvania, 1935), pp. 3-6.

10. John B. Rae, "Engineers Are People," *Technology and Culture*, 16, 3 (July 1975), p. 416.

11. Calhoun, op. cit., pp. 164-67.

12. *Dictionary of American Biography* op. cit., vol. 11, p. 380.

13. Ibid., vol. 12, pp. 152-53, and vol. 20, p. 72.

14. John B. Rae, "Scottish Engineers in America," Newcomen Society *Transactions*, 49 (1977-78), p. 127.

15. Neil Fitz-Simons (ed.) *The Reminiscences of John B. Jervis, Engineer of the Old Croton* (Syracuse: Syracuse University Press, 1971)

16. John H. White, *A History of the American Locomotive: Its Development, 1830-1880* (New York: Dover, 1979), pp. 167-69.

17 Morison, op. cit., pp. 65-66.

18 Gene D. Lewis, *Charles Ellet, Jr.: The Engineer as Individualist* (Urbana: University of Illinois Press, 1968)

19 Darwin Stapleton, "Moncure Robinson," in Barbara Benson (ed.), *Benjamin Henry Latrobe and Moncure Robinson: The Engineer as Agent of Technological Transfer* (Wilmington: Eleutherian Mills-Hagley Foundation, 1975), pp. 33-60.

20 Ibid., p. 49.

21 *Biographical Dictionary of American Civil Engineers*, op. cit., pp. 77-78.

22 Edward C. Carter II, "Benjamin Henry Latrobe," in Benson, op. cit., pp. 11-27.

23 Totals for both tables do not equal 100 percent due to missing information.

24 David Stevenson, *The Civil Engineer of North America* (London: J. Weale, 1838), pp. 192-93. Stevenson was a partner in an Edinburgh firm with his brother, Robert Stevenson, whose son, Robert Louis Stevenson, was also trained as an engineer, but left engineering to embark on a career as a writer, with highly successful results.

25 *Biographical Dictionary of American Civil Engineers*, op. cit., p. 36.

26 Ibid., p. 37.

27 James H. Hall and G.W. Richardson, "George H. Corliss," *Mechanical Engineering* (1933), pp. 403-409.

28 Hiram F. Mills, "James Bicheno Francis," read to the Corporation of the Institute of Technology, 11 December 1892; reproduced in L.S. Bryant and J.B. Rae, *Lowell: An Early American Industrial Community* (Cambridge, Massachusetts: MIT Press, 1950)

29 James Carvill, *Famous Names in Engineering* (London: Butterworths, 1981), p. 30.

30 John F. Stover, *American Railroads* (Chicago: University of Chicago Press, 1961), p. 159.

31 For a survey and analysis of the development of metal-working in nineteenth-century America, see David Hounshell, *From the American System to Mass Production* (Baltimore: Johns Hopkins University Press, 1984), pp. 1-187.

32 Joseph W. Roe, *English and American Tool Builders* (New York: McGraw-Hill, 1926), pp. 247-269.

33 Bruce Sinclair, *Philadelphia's Philosopher Mechanics: A History of the Franklin Institute 1824-1865* (Baltimore: The Johns Hopkins University Press, 1974), pp. 311-16.

Chapter 8

Into the Twentieth Century: Engineers Join the Organizational Revolution

Much of what has been written in the previous chapters has centered on the accomplishments of individual engineers. Throughout the centuries leading up to our own, engineering tended to be an individualistic endeavor, dominated by the heroic figures that we have met in the pages above. But during the nineteenth century the scope and scale of engineering began to widen as the development of large organizations gave a powerful stimulus to the training and employment of large numbers of engineers. Toward the end of this century the engineer's working life was beginning to change significantly as engineering was being accommodated to the requirements of large organizations. In this chapter, and in succeeding ones as well, the narrative will take up some new issues as the organizational context of engineering is brought into focus. Although we will continue to encounter individual engineers and their achievements, it will also be necessary to give more consideration to engineering as a group endeavor.

In the United States the association between the growth of organizations and the expansion of engineering ranks began to manifest itself at an early date. As was noted in the last chapter, during the second decade of the nineteenth century the United States was a predominantly agrarian nation, with a mere handful of people who could be counted as engineers. When work on the Erie canal began it marked the initiation of the largest single enterprise the country had ever seen, and not coincidentally it led to a doubling of the number of practicing engineers.[1] The Erie Canal's stimulation of the engineering trade gave some indication of what was to come. To an increasing degree throughout the nineteenth and twentieth centuries, the occupation of engineer was intertwined with the growth and development of large organizations.

To be sure, engineers have always been closely tied to large-scale enterprises. From the pyramids of ancient Egypt to the Erie Canal of early nineteenth century America, much of the work of engineers was directed at the design, construction, and operation of devices and structures that required a great deal of concerted human effort. Yet while they were part of an organized effort, most of the engineers of the pre-modern period were not permanent employees of the enterprises for whom they worked. Rather, they were self-employed designers who sold their services to the organizations that needed them. When their work was completed they moved on to another client.

In some sectors of the trade, most notably civil engineering, this age-old practice continued into the industrial era, where it produced some of the era's most innovative and impressive works of engineering.[2] The individualistic genius of the independent engineer is exemplified by one of the proudest creations of late nineteenth century America, the Brooklyn Bridge. Its designer, John Augustus Roebling (1806-1869), was born in Germany and received a civil engineering degree from the Royal Polytechnical School in Berlin. After emigrating to the United States he set up a successful iron cable company. This enterprise was a natural complement to his interest in suspension bridges, and in 1844 he built his first suspension bridge after winning a competition to design a structure to carry a canal across the Allegheny River. In the following year he built another suspension bridge, this one to move vehicular traffic over the Monongahela River in Pittsburgh. During the years following, his design of bridges at Niagara Falls and Cincinnati prepared him for his greatest work. But while surveying for the Brooklyn Bridge in 1869 his foot was crushed in a collision at a ferry slip, and he died of tetanus in less than a month. His talented son Washington Roebling 1837-1927) took over, but he too suffered an on-the-job accident, a caisson fire that left him an invalid. Nevertheless, he continued to oversee the the construction of the bridge until its completion in 1883.[3]

While the Roeblings pioneered the use of long-span suspension bridges, Othmar Ammann (1879-1965) demonstrated the spectacular capabilities of the basic design. After emigrating to the United States from his native Switzerland, Ammann spent eleven years working with another renowned bridge designer, Gustav Lindenthal. Striking out on his own in 1923, he submitted his plans for a bridge to cross the Hudson River in New York City. Not only were his plans accepted, in the following year he was named chief engineer of the Port of New York Authority. He went on to design two impressive cantilever truss bridges, the Goethals Bridge and the Outerbridge Crossing, followed by the longest suspension and arch bridges of their time, the George Washington and Bayonne bridges, respectively.

Ammann's independent position allowed the free reign of his creative talents. Ammann was motivated by aesthetic considerations at least as much as structural ones, and he endeavored to build suspension bridges with especially thin decks, thereby producing an pleasing visual contrast between the deck and the massive support towers. This was masterfully achieved with the George Washington bridge, but a few years later the collapse of another engineer's product, the Tacoma Narrows Bridge, forced a great deal of re-evaluation of the structural integrity of thin decks, and ultimately motivated Ammann's most important structural innovation, the use of a tubular framework for the deck of the Verrazano Narrows Bridge, completed in 1965.

Ammann pushed established concepts to their limits, but another Swiss engineer, Robert Maillart (1872-1940), was a true innovator in his use of a new structural material for the consturction of bridges of unsurpassed elegance and economy. The material was concrete reinforced with iron, first used by Francois Hennebique. Hennebique had demonstrated considerable entrepreneurial drive, and by the beginning of the 20th century he was overseeing a large international network of franchised firms employing the *System Hennebique*. Still, it took the individual genius of Maillart to demonstrate the potential of reinforced concrete to blend functional superiority with great aesthetic appeal. Maillart headed his own design firm, through which he designed forty-seven major bridges, all but three of which are still in use. He did not create a commercial empire like Hennebique's, but his independence allowed him to pursue his vision while free from an employer's dictates.

Representatives of an earlier age of engineering, the careers of these renowned bridge designers were not typical of the direction in which engineering was going, for economic and technological advance was dramatically changing the working environment of the typical engineer. By the end of the nineteenth century, the trend had become unmistakable. Not only was the industrial economy expanding rapidly, so was the size of enterprises. The largest American firms in 1870 were textile factories, iron and steel mills, and agricultural implement works, none of which had more than a few hundred employees. Within a few decades the situation changed markedly. By 1900 more than a thousand factories employed between 500 and 1000 workers, and 443 of them had more than 1000 workers.[4] Some giant enterprises such as Carnegie Steel and Baldwin Locomotive had up to 10,000 employees.[5] The age of the giant enterprise had begun.

This trend continued through the twentieth century as the expansion of markets, the development of transportation networks, and technological advance produced immense enterprises that dominated the industrial economy. In the United States today, 200 manufacturing firms hold 60 percent of industrial assets, while 500 manufacturing firms account for 70 percent of industrial sales and 75 percent of industrial employment. At the same time, federal, state, and local government has expanded to the point where today it collectively accounts for more than 20 percent of the U.S. gross national product, and about the same percentage of the workforce.

The engineering occupation followed these trends. Fewer than 5 percent of engineers are now self-employed.[6] Not only do most engineers serve as employees, they generally do so in large organizations. According to one study conducted in the mid-1960s by the American Society for Engineering Education, three-quarters of American engineers were employed by firms with more than 1000 employees,[7] while 35 percent could be found in firms with more than 10,000 employees.[8] In the early 1990s the largest 100 U.S. companies employed more than 30 percent of the nation's engineers, and 50 percent of those in manufacturing. The largest employer of engineers, General Electric, had 35,000, while the second largest, General Motors, employed 30,000.[9] This phenomenon is not unique to the United States; in most industrial countries, between thirty and forty percent of university graduate engineers are employed by organizations with 5000 employees or more.[10]

The Growth of Engineering Specializations

While the large firm was taking over as the typical occupational environment of engineers, engineering work was becoming increasingly specialized. As we have seen, many nineteenth-century engineers engaged in both civil and mechanical engineering, but by the end of the century the two had become distinct fields. At the same time, two new specialties were emerging, chemical and electrical engineering. In 1930, this quartet still accounted for 90 percent of all engineers, but toward the end of the 1980s only 64 percent of engineers were working in these fields.[11] Many engineers now specialize in areas which had been virtually unknown a generation ago; geophysical, nuclear, materials, industrial, aerospace, and environmental engineering are now recognized as distinct areas of engineering practice.

As engineering has fragmented into a large number of specialties, most engineers now contribute only a small segment to a single project. Working as part of a large team, many engineers now contribute to sizeable collective efforts. The design and development of a single modern transport plane such as Lockheed's C5A, for example, occupied no fewer than 2000 engineers,[12] and many large projects routinely require the services of hundreds of engineers.

The Industrial Research Laboratory

Engineering has always been an activity distinct from science. Still, there is no denying that a vast amount of engineering practice has emerged from scientific knowledge. Pushed by the commercial requirements of business organizations, science in the twentieth century has joined into a close (and some would say unholy) alliance with technology. Equally important, while science and technology have grown closer, the process of invention has been systematized. A considerable amount of engineering work is now being done in a type of institution that emerged about a century ago: the industrial research laboratory.

The prototype institution was Thomas Edison's laboratory in Menlo Park, New Jersey. Although the folklore surrounding Edison dwells on the accomplishments of a solitary genius, Edison in fact owed much of his success to the talented research team that he had assembled. Their individual skills were nurtured and conjoined in an extensive research facility that included a chemical laboratory, machine and carpentry shops, and an electrical testing department. In this compound Edison and his associates pursued a number of epochal inventions, most notably a practical filament for incandescent lamps, along with all of the apparatus necessary for a commercially viable electrical lighting system.[13]

General Electric, the company that absorbed Edison's firm, took the idea of systematic research and development to a higher level. Under the directorship of Willis Whitney (1868-1958), the GE lab combined the talents of scientists, engineers, and skilled craftsmen. At the GE lab the distinction between science, applied science, and engineering was hard to draw as teams of researchers tackled practical problems that required both scientific and engineering talents. The most notable of the lab's early projects were the development of the tungsten filament and the use of inert gases that together made for much longer-lasting light bulbs.[14] Of almost equal importance, the GE labs came to symbolize the technological virtuosity of a large corporation that was able to produce a continual stream of new products and processes through the organized use of advanced science and engineering.

Other companies followed a similar strategy during the early years of the 20th century, most notably Kodak, Dupont, and Standard Oil. By the 1920s, General Motors was sponsoring a well-orchestrated engineering research program under the leadership of Charles Kettering (1876-1958). A practical man best known for his invention of the self-starter and his role in the development of tetraethyl lead as an octane booster, Kettering nonetheless valued sustained research efforts as the basis of future engineering applications. As he put it, "The general tendency, looking at any research problem, is to make it too ingrown. By that I mean referring to research as the solving of manufacturing difficulties only and not to the bigger problems of the future."[15] Guided by Kettering's vision, General Motors pursued research projects that eventually emerged as new products, most notably the automatic transmission, the two-stroke diesel engine, and the diesel-electric locomotive.

The greatest example of the large-scale industrial research has been the Bell Laboratories. At their height, the Bell Labs employed 14,000 people, 700 of whom held doctoral degrees.[16] It was in this setting that the combined efforts of John Bardeen, Walter Brattain, and William Shockley (a theoretical physicist, an experimental physicist, and an electrical engineer, respectively) produced one of the century's most important inventions, the transistor. In this case, and many others like it, engineers were able to do their best work while working closely with scientists, both groups drawing support from the resources that only a large-scale organization could muster.

As the growth of large-scale private enterprises produced a radically different climate for engineering work, the expansion of government reinforced this trend. Especially in the post-World War II era, growing numbers of engineers have been directly employed by the government or by firms dependent on government contracts. The federal government itself employed 106,195 engineers in 1986, more than 4 percent of all working engineers.[17] More important, about half of research and development work conducted in the United States is supported by government funds. Much of this expanded governmental role has been due to the vast influence of the military sector on the course of technological development and application. The military is of course hardly an alien intruder into engineering, as previous chapters have shown. What makes the post-war era unique, however, is the great increase in both the scope and the technological complexity of warfare. Under such circumstances the engineer working as part of a larger aggregate has become an essential element of modern warfare. The engineer's products gave us modern warfare; at the same time modern warfare made a strong contribution to the practice of modern engineering.

The results of the militarization of engineering are evident. Depending on how the count is made, the Department of Defense either directly or indirectly supports between a fifth and a third of America's scientists and engineers, while military needs strongly influence the expenditure of funds for basic and applied research. Motivated to a great extent by military considerations, the federal government has heavily supported the research and development efforts of a few industries; in aerospace it supplies 80 percent of R&D funding, while for for electrical equipment and telecommunications it provides nearly half.[18]

The immense importance of the defense industry in the sponsorship of engineering work strongly reinforces the tendency for engineers to work in large organizations. During the 1970s, for example, one-third of the Pentagon's research, development, testing, and evaluation funds went to only eight firms. These firms also had between 14 and 78 percent of their independent research programs reimbursed by the government.[19]

The influence of the military is also demonstrated in the considerable fluctuation of post-war engineering employment. When defense expenditures have been at a high level, engineering employment has followed suit; when defense appropriations have been cut back, as has happened in the early 1990s, large numbers of engineers have found themselves out of work. Accordingly, as one government report noted, "government programs and policies greatly influenced the boom and bust cycles of science and engineering labor markets over the past several decades."[20]

The other half of R&D funds, those provided by the private sector, are also disproportionately disbursed by a relatively small number of firms. In the U.S., it has been estimated that 10,000 firms perform R&D work, but 20 of them collectively account for 50 percent of the total, and only four account for 20 percent.[21] We can see in these statistics a reflection of the earlier figures that showed the majority of engineers being employed by a relatively small number of large firms.

The Engineer and the Organization: Conflict and Concord

Most of the engineers who work in typical organizational settings are anonymous, their work known only to their colleagues. But this is not always the case. Human genius can manifest itself in a variety of environments, and even in large organizations it is possible to find individual engineers whose contributions make them stand out. In some cases their genius has been recognized and rewarded; other times it has not, sometimes with tragic results. The careers of the three men who appear below exemplify different aspects of the relationship between engineers and large-scale organizations.

William Stanley (1858-1916) was a largely self-taught engineer who first worked for two pioneers of the electrical industry, Hiram Maxim and George Westinghouse.[22] His worth to the latter's firm soon became evident: in his first year-and-a-half of employment Stanley was granted patents for fourteen inventions. Uncomfortable with the the workaday world of Westinghouse's engineering program, he was allowed to set up his own research facility with the stipulation that any inventions produced there would be for the exclusive use of the company. It was here that he produced his most important innovation, a technique for mass-producing transformers. This helped Westinghouse to take the lead in the use of alternating current, which proved to be essential for the long-range transmission of electricity. But Westinghouse was not convinced that Stanley had produced anything fundamental, and the company's founder refused to give Stanley what the latter felt was his due.

Stanley went on to form his own electrical company, which successfully developed transmission systems for hydro-electric power in the western United States. Ironically, however, his company was sued for patent infringement by Westinghouse, which held the rights to a patent he had earlier signed over to them. Even so, Stanley did not suffer grievously. General Electric paid him $32,000 for the 22 patents he had acquired after leaving Westinghouse, and hired him at $12,000 a year to work in his own laboratory (which was financially supported by GE) with the stipulation that he spend half of his time working on problems sent to him by GE. With the remainder of his time he ran his own thermos bottle company, making use of manufacturing innovations he had created. But while Stanley enjoyed a considerable measure of financial prosperity, he was always annoyed that his contributions never had been properly recognized and rightfully attributed to him.

A much sadder story is that of Edwin Armstrong (1890-1954). A graduate of Columbia University's electrical engineering program, Armstrong's first triumph was the regenerative (or feedback) circuit that greatly increased the range at which radio signals could be received. He then went on to develop the superheterodyne receiver, which used the electrical mixing of frequencies to amplify incoming signals more effectively. In 1933 Armstrong demonstrated a way to largely eliminate static—one of the most vexing problems of radio—through the use of frequency modulation. In so doing he laid the foundations of the FM broadcasting industry. Armstrong financed his inquiries into FM radio through the use of royalty money awarded for his earlier inventions. As FM became a commercial reality in the 1940s, RCA and other firms made ample use of his ideas, leading Armstrong to file a patent infringement lawsuit. His own financial resources were no match for those of RCA, and the extended process of litigation left him depressed and exhausted. A broken man, he committed suicide at age 63. Ironically, after his death the courts found in his favor, and over $1,000,000 was awarded to his estate. Although he largely worked as an independent inventor, he was hardly invulnerable when confronting a large corporation, and he was defeated by its overwhelming financial and legal resources.[23]

To be sure, not all engineers have had such unpleasant relationships with corporate organizations. Some, as exemplified by Clarence "Kelly" Johnson (1910-1990), have been able to reconcile their creativity as engineers with the demands of a large organization.[24] Johnson came to Lockheed Aircraft in 1933, fresh out of the University of Michigan, where as a student he had done wind tunnel tests on Lockheed's new commercial airliner, the Electra. Although he had just arrived, he wasted little time in informing his superiors that the craft was unstable, and would require a redesign of the tail assembly. Far from taking umbrage at the advice of a novice engineer, Lockheed soon had him working on a redesign of the tail. After a considerable amount of wind tunnel work, he added controllable plates to the horizontal stabilizer and redesigned the tail, giving the the airplane its characteristic two-fin configuration.

Johnson rose quickly to the top rank of Lockheed's engineering staff, becoming chief research engineer in 1938. In the middle of 1943 he began work on the the P-80, the Army Air Corps second jet fighter and the first one to be used in combat. Under Johnson's direction, the P-80 was designed by a staff of 120, including 23 engineers, and flew 143 days after the original procurement order was signed. In the course of designing and producing the prototype P-80, Johnson created his most important organizational innovation: the famed Lockheed "Skunk Works." The Skunk Works came to be the renowned birthplace of a host of advanced aircraft, most notably the supersonic F-104 fighter, the U-2 high-altitude reconnaissance plane, and its successor, the SR-71. So advanced was the SR-71 that 26 years after its first flight it marked its retirement by setting a coast-to-coast speed record: Los Angeles to Washington in 68 minutes.

Johnson's Skunk Works encompassed a wide variety of operations that included everything from initial design studies to the construction and test of prototype aircraft. It produced results rapidly because it was small in size and simple in its organizational structure—a sharp contrast with its parent organization. As Johnson summarized its operation:[25]

> The ability to make immediate decisions and put them into rapid effect is basic to our successful operation. Working with a limited number of especially capable and responsible people is another requirement. Reducing reports and other paperwork to a minimum, and including the entire force in the project, stage by stage, for an overall high morale are other basics. With small groups of good people you can work quickly and keep close control over every aspect of the project.

Making and implementing decisions quickly required a considerable concentration of managerial authority: "The Skunk Works manager must be delegated complete control of his problem in all aspects. He should report to a division president or higher."[26] The investment of a single manager with this kind of authority contrasts with the contemporary tendency toward committee design and its dilution of the individual's role. The result of the latter, according to Johnson, is that "Nothing very stupid will result, but nothing brilliant either. And it's in the brilliant concept that a major advance is achieved."[27]

Kelly Johnson occupied the summit of engineering at Lockheed, exercising his influence over aircraft design in ways that few other engineers have enjoyed. From the organizational bastion of the Skunk Works he was able to initiate and guide projects while at the same time taking the lead role in their design. Although his accomplishments as an aircraft engineer were significant, even inspiring, his career was hardly typical of the great majority of engineers. And this is explicitly recognized in the subtitle of his autobiography: More Than My Share of It All.

The Ascent of Japanese Engineering Exemplified

A few years before Kelly Johnson went to work for Lockheed, a Japanese engineer was beginning a career in aeronautical engineering that was to produce equally signficant accomplishments. This in itself showed how the world of engineering was changing. No society has had a monopoly of engineering talent, but despite the technological achievements of the non-Western world, during the 19th and much of the 20th centuries the process of industrialization was largely initiated and developed to Europe and North America. Consequently, the most advanced engineering work was found in these regions. This situation began to change toward the end of century as Japan began to embark upon an ambitious program of industrial development.

Japan's industrialization, coupled with an increasingly aggressive military posture, provided the setting for the cultivation of native engineering capabilities. The rise of Japanese engineering to a position of eminence is exemplified by the career of Jiro Horikoshi.[28] Born into a rural family in 1903, Horikoshi entered Tokyo University's Department of Aeronautics in 1923. Upon the completion of his training three years later, he joined the Nagoya Aircraft Manufacturing Plant of the Mitsubishi Internal Combustion Engine Company (later Mitsubishi Heavy Industries). In the early 1930s Horikoshi designed a number of carrier-based monoplane fighters that largely followed Western practice. Later in the decade he produced the supreme accomplishment of the Japanese aircraft industry, the Type 00 fighter, the famous "Zero" that wreaked havoc in the skies over the Pacific during World War II. Although the Zero had first seen action against China, it came as a complete surprise to the Allies, who up to Pearl Harbor smugly believed that the Japanese aircraft industry could only produce pale imitations of their own craft. In this they were sadly mistaken, for during the early years of the war the Zero proved itself more than a match for British and American fighters because of its great range, rate of climb, and maneuverability. Allied pilots learned that it was almost suicidal to engage in one-to-one combat; Zeroes could be successfully attacked only when outnumbered and their superior performance thereby offset.

In designing the Zero, Horikoshi did not employ any radical departures from contemporary practice. His primary goal was keeping weight to a minimum, and to do this he consciously lowered the safety factor for those parts that could withstand bending loads right up to the maximum, while retaining high safety factors for components that would begin to fail well before this point.[29] Much of the Zero's success could also be attributed to Horikoshi's application of clever details. He designed the wingtips with a slight downward twist at the leading edge to prevent them from stalling prematurely. He also solved the longstanding problem of control response being affected by the speed of an aircraft by designing the control cables with a predetermined amount of stretch; in this way equal movements of the stick would produce the same amount of aircraft movement at both high and low speeds.[30]

The Working Life of Engineers

The expanded size of enterprises and their reliance on increasingly sophisticated technologies have heavily contributed to a great expansion in the number of engineers. But at the same time, the new occupational environment produced by large enterprises has fundamentally altered the working lives of engineers. As technological knowledge and application have dramatically expanded, the last hundred years or so have been the era of the engineer, but paradoxically the engineer's status may be more uncertain and confused than it was a century ago.

As noted earlier, most of today's engineers are employees of large organizations. This has had important consequences for the occupational life of engineers. As organizations grow in size they change in structure. What happens, in essence, is bureaucratization. A multitude of specialized occupational roles have to be coordinated and integrated into a coherent system. The primary means of doing this is through the application of rules and regulations, and the exercise of authority through hierarchical chains of command.

What a bureaucracy requires of its employees does not always tally with what engineers want in their work environment. Bureaucratic organizations require loyalty and obedience first; creativity may be of less importance. Indeed, it may even be an subversive element undermining the comfortable routines and procedures of these organizations. At the same time, the hierarchical authority typical of bureaucracies may conflict with authority based on technical expertise. This raises the possibility of an engineer's expert judgment being overridden by less-qualified officials who occupy a higher position within the organization.

The apparent conflict between engineers and the organizational structures in which they work has led some social theorists to predict a major conflict between the two, one that would ultimately result in the political and economic ascendancy of the engineers (see Chapter 9). But in fact this is not the way things have turned out. For the most part, engineers have accepted the necessity of contemporary organizational structures, including the exercise of hierarchical authority and the pursuit of organizational goals not necessarily of their own making.

Although sociological studies of engineers' work are rare, the few that have been done largely concur about the working life of engineers and the strictures that they are subject to. While a bureaucratically organized work environment is necessarily hierarchical, engineers as a group are not subject to the same degree of control as other employees. Most engineers work in "professional bureaucracies" in which command structures are flexible, giving individual engineers a considerable amount of discretion.[31] Also, while there are usually some elements of control based on hierarchical position, they are not deeply resented, for the exercise of managerial authority tends to be seen as coordination rather than supervision.[32] Finally, engineers implicitly accept hierarchical arrangements in that they consider their immediate supervisors to be the most important in judging their performance, and not their fellow engineers.[33]

At the same time, however, there appears to be considerable frustration over the technical ignorance of many supervisors. Engineers in these circumstances are likely to feel that they are not given the opportunity to demonstrate their highest skills, resulting in generally negative feelings about the work being done.[34] As a result, when engineers think of workplace democracy (should they be concerned with it at all) they see it not as an end in itself, but as a means of promoting greater efficiency through an increase in technical rationality.[35]

In addition to accepting managerial authority on principle, engineers seem to agree that the ultimate purpose of their craft is contributing to the financial health of the company that employs them. Very few engineers believe in engineering-for-the-sake-of-engineering or that their work can be legitimated solely through its technical excellence.[36] Few engineers would dissent with Edwin Layton: "Engineering is a scientific profession, but the test of an engineer's work lies not in the laboratory, but in the marketplace."[37] Under these circumstances, engineers are not likely to make technical virtuosity an end in itself. Engineers cannot evaluate their work according to purely technical criteria; financial considerations are never far from view. In the words of a distinguished British engineer: "The discipline of finance must ultimately control all engineering; the economists and accountants find their place in the industrial management team because technology, time, and money are inseparable in industry, and money is the only yardstick for the ultimate measurement of industrial success."[38]

The close connection between engineering excellence and financial success is exemplified by Charles Kettering's working life. Kettering labored under no illusions about the engineer's role. He was firmly convinced that what made for a successful engineer was not simply technical virtuosity, but an ability to contribute to his firm's bottom line. The driving force behind two pioneer industrial laboratories, first at National Cash Register and then at General Motors, Kettering emphasized that "the man responsible for the laboratory must never lose sight of the fact that research done with corporate funds must justify itself economically."[39] This in turn meant inventing and developing products that had a strong market: "It is all very well to make discoveries in the laboratories, but if these do not find their way into processes and products, then the research was wasted."[40] And, as far as Kettering was concerned, the best way of orienting his work to the needs of the market was to stay in close touch with those who understood consumers and their requirements: "I didn't hang around much with other inventors or the executive fellows. I lived with the sales gang. They had some real notion of what people wanted."[41]

Kettering's ideas are not far from those of most engineers. The constraints of finances and the market are not seen as unpleasant obstacles to their demonstrating technological virtuosity. Rather, they are seen as being natural and inevitable, and certainly not alien concerns held only by management.[42] Engineers for the most part do not subscribe to an independent "logic of technology" that stands opposed to the logic of profit-making; rather, technology is seen as serving the ethos of business that permeates the working environment.[43] As one engineer put it, "Everyone wants to design the ultimate machine, but you can't fly in the face of economic reality. When I decide what to do on my own, I do it on the basis of cost considerations; and I can find something just as challenging. They aren't mutually exclusive."[44]

This realization of engineering's economic dimension also softens resistance to managerial guidance. When engineers are subject to managerial constraints, they tend to see them not as alien impositions, but as the reasonable requirements of a firm competing in the marketplace.[45] Engineering work has to adhere to the budgetary standards set by management. And, for the most part, engineers do not examine these standards with a critical eye, but accept them as an inherent part of engineering.[46]

Although the work of engineers has increasingly been subject to bureaucratic controls and precise financial calculations, the work of engineers has not been subject to the de-skilling suffered by some other occupations as they became subject to managerial controls.[47] Most firms value the skills of their engineers, and grant them considerable latitude in the way they put them to use. Above all, engineers are generally viewed as permanent, trustworthy members of the organization. Unlike many other workers, they are not treated as replaceable parts to be used and discarded according to the immediate needs of management.[48] As a result, engineers are given the opportunity to use many of the skills that they bring with them and to acquire new ones. Especially for engineers with advanced degrees who have acquired several years' worth of job experience, there are many opportunities to do "cutting-edge" work that requires high levels of skill.[49]

But if engineers as an occupational group have not been subject to widespread deskilling, there are still many engineers who could do quite a bit more with the skills that they possess. Although engineers constantly bemoan the difficulty of keeping up with rapidly advancing technical knowledge, a significant portion of engineers faces a very different issue. For them, the problem is not keeping up with new knowledge but using what they already have. Lacking an opportunity to do what they are capable of doing, these engineers are subject to a progressive atrophy of their skills. In the summation of one sociologist, "The problem is more one of technical amnesia than of technical senility."[50]

There is no escaping the fact that a great deal of engineering work is quite routine; it consists of the performance of rigidly structured tasks that entail little in the way of novelty. These routine tasks provide few opportunities to expand expertise, or to influence a firm's technical policies.[51] Under these circumstances, many engineers are deprived of the opportunity to make full use of the skills that their education has given them. Equally important, for engineers ensnared in routine work there are few opportunities to demonstrate their craftsmanship—one of the primary sources of job satisfaction.[52]

Many engineers would welcome the opportunity to improve their technical capabilities. Even so, efforts to increase job satisfaction by promoting and recognizing technical accomplishments have not been successful.[53] Although craftsmanship may be the dominant source of intrinsic satisfaction, engineers are strongly attracted to positions in management. Indeed, for the majority of engineers, the prospect of a future position in management is one of the strongest sources of motivation for entering engineering in the first place, and for doing good work thereafter. Accordingly, attempts by firms to establish "dual ladders" (one for those aspiring to managerial positions, and one for those oriented towards achievement in the technical realm) have not found a receptive audience within engineering ranks. Far from being a viable alternative to the managerial ladder, the technical ladder is almost always viewed as a device used to maintain commitment to the organization for those who have failed to move into the ranks of management.[54]

The weak appeal of technical prowess as an end itself has important implications for the professional status of engineering. This will be explored in Chapter 11, but before we do so, we need to consider the managerial ladder, its history, and the extent to which engineers have succeeded in moving into management. These will be the topics of the next chapter.

Notes

1 Daniel C. Calhoun, *The American Civil Engineer* (Cambridge, Massachusetts: MIT Press, 1960), p. 30.

2 The following section draws heavily on David P. Billington, *The Tower and the Bridge: The New Art of Structural Engineering* (Princeton: Princeton University Press, 1985).

3 Hamilton Schuler, *The Roeblings* (Princeton, New Jersey: Princeton University Press, 1931).

4 Daniel Nelson, *Managers and Workers: Origins of the New Factory System in the United States 1880-1920* (Madison: University of Wisconsin Press, 1975), p. 6.

5 Ibid., p. 7.

6 Samuel Florman, *The Existential Pleasures of Engineering* (New York: St. Martin's Press, 1976), p. 22.

7 William K. LeBold, Robert Perrucci, and Warren Howland, "The Engineer in Industry and Government," *American Society for Engineering Education Proceedings*, 1966, p. 243. Cited in Richard Ritti, *The Engineer in the Industrial Corporation* (New York and London: Columbia University Press, 1971), p. 57.

8 Robert Perrucci and Joel Gerstl, *Profession without Community: Engineers in American Society* (New York: Random House, 1969), p. 2.

9 *Scientific-Engineering-Technical Manpower Comments*, April-May 1992, p. 3.

10 Wouter Van den Berghe, *Engineering Manpower: A Comparative Study of the Employment of Graduate Engineers in the Western World* (Paris: UNESCO, 1986), p. 114.

11 Computed fron *Scientific-Engineering-Technical Manpower Comments*, March 1988, p. 5.

12 David Sawers and Ronald E. Miller, *The Technical Development of Modern Aviation* (New York: Prager, 1970), p. 276.

13 Thomas Parke Hughes, "Thomas Alva Edison and the Rise of Electricity," in Carroll Pursell, Jr. (ed.), *Technology in America: A History of Indi-*

viduals and Ideas (Cambridge, Massachusetts: MIT Press, 1981), pp. 119-24.

14 Elting E. Morison, *From Know-How to Nowhere: The Development of American Technology* (New York: New American Library, 1977), pp. 107-31.

15 Stuart W. Leslie, *Boss Kettering* (New York: Columbia University Press, 1983), p. 117.

16 Morison, op. cit., p. 96.

17 *Federal Science and Engineering* (Washington, D.C.: National Science Foundation, 1986)

18 Colin Norman, *The God that Limps: Science and Technology in the Eighties* (New York: Norton, 1981), p. 90.

19 William D. Hartung, et al., *The Economic Consequences of a Nuclear Freeze* (New York: Council on Economic Priorities, 1984), p. 68.

20 Quoted in James Botkin, et al., *Global Stakes: The Future of High Technology in America*, (New York: Penguin Books, 1984), p. 81.

21 Norman, op. cit., p. 91.

22 George Wise, "William Stanley's Search for Immortality," *American Heritage of Invention and Technology*, 4, 1 (Spring/Summer 1988)

23 Lawrence Lessing, *Man of High Fidelity: Edwin Howard Armstrong* (Philadelphia: J.P. Lippincott, 1956)

24 This section is based on Clarence L. "Kelly" Johnson with Maggie Smith, *Kelly—More than My Share of It All* (Washington, D.C.: Smithsonian Institution Press, 1985).

25 Ibid., p. 160.

26 Ibid., p. 170.

27 Ibid., p. 164.

28 Jiro Horikoshi, *Eagles of Mitsubishi: The Story of the Zero Fighter* translated by Shojiro Shindo and Harold N. Wantiez. (Seattle: University of Washington Press, 1981)

29 Ibid., pp. 37-39.

30 Ibid., pp. 44-46 and 73-80.

31 Robert Zussman, *Mechanics of the Middle Class: Work and Politics among American Engineers* (Berkeley and Los Angeles: University of California Press, 1985), p. 110; Stephen Crawford, *Technical Workers in Advanced*

Society: The Work, Careers, and Politics of French Engineers (Cambridge: Cambridge University Press, 1989), pp. 100-106.

32 Peter Whalley, *The Social Production of Technical Work: The Case of British Engineers* (Albany: State University of New York Press, 1986), p. 130.

33 Perrucci and Gerstl, op. cit., p. 119.

34 Joel E. Gerstl and Stanley P. Hutton, *The Anatomy of a Profession: A Study of Mechanical Engineers in Britain* (London: Tavistock, 1966), p. 120.

35 Zussman, op. cit., p. 118.

36 Whalley, op. cit., p. 133; Crawford, op. cit., p. 129.

37 Edwin T. Layton, Jr., *The Revolt of the Engineers: Social Responsibility and the American Engineering Profession* (Baltimore and London: Johns Hopkins University Press, 1986), p. 1.

38 Christopher Hinton, *Engineers and Engineering* (Oxford: Oxford University Press, 1970), p. 46.

39 Leslie, op. cit., p. 208.

40 Ibid., pp. 184-85.

41 Ibid., p. 36.

42 Whalley, op. cit., p. 137.

43 Zussman, op. cit., p. 224.

44 Ibid., p. 120.

45 Whalley, op. cit., p. 149.

46 Zussman, op. cit., p. 223.

47 The most influential statement of the "deskilling" thesis is Harry Braverman, *Labor and Monopoly Capital: The Degradation of Work in the Twentieth Century* (New York and London: Monthly Review Press, 1974). An argument that engineers have suffered from capitalist-induced deskilling is presented in Mike Cooley, *Architect or Bee?* (Boston: South End Press, 1982).

48 Whalley, op. cit., pp. 87-90; Crawford, op. cit., p. 230.

49 Perrucci and Gerstl, op. cit., pp. 135-36.

50 R. Richard Ritti, *The Engineer in the Industrial Corporation* (New York and London: Columbia University Press, 1971), pp. 219-20.

51 Ibid., p. 42.

52 Gerstl and Hutton, op. cit., p. 119.

53 Fred H. Goldner and R.R. Ritti, "Professionalization as Career Immobility," *American Journal of Sociology*, 72 (March 1976) pp. 489-502.

54 Ibid., p. 490.

Chapter 9

Engineers as Managers

A career in engineering can offer substantial financial and psychic rewards. But for many it has been only the first stage of a career. Engineers have moved into a variety of occupations; Jimmy Carter became President of the United States, while Neville Shute went on to be a best-selling novelist. The career changes of most engineers have been more modest, although of considerable individual and social significance. If there is one commonality in the occupational lives of large numbers of engineers, it is that they have moved from engineering into management.

The progression from engineer to manager has had important consequences for both occupations. On the one hand, the attitudes, values and working methods of the engineer have strongly influenced the practice of management during much of the 20th century. Engineers have made major contributions to the evolution of management theory and practice, thereby extending the influence of engineering into many new areas. At the same time, however, the prospect of movement into management has undermined the effort to make engineering's status equal to that of the estalished professions. All in all, the close association between management and engineering has been a complex one, with elements of both symbiosis and conflict.

Engineers, Owners, and Managers

Although the movement from engineering into management has been most evident during the 20th century, it has ample historical precedent. Many of the engineers described in earlier chapters necessarily took on managerial functions as they oversaw the construction and operation of the projects that they had designed. Of equal or greater importance, significant numbers of engineers assumed managerial responsibilities as proprietors of their own firms. For many nineteenth century engineers, the role of engineer, manager, and owner could scarcely be distinguished; practicing engineers were as likely to be managers of their own firms as they were providers of technical skills to an outside clientele.[1]

The organizational revolution that began to gather steam towards the end of the nineteenth century did not put an end to the engineer-proprietor. On the contrary, a few engineers built large organizations on the foundations of an early career in engineering. One outstanding example is the firm that was founded by Charles A. Stone (1867-1941) and Edwin S. Webster (1867-1950). The two had studied electrical engineering together at the Massachusetts Institute of Technology (MIT), and after graduation set themselves up as consultants to the emerging electrical industry in the Boston area. Their first tasks were mundane ones like testing wiring and electrical appliances for insurance companies. But within two years they were engaged in the more challenging venture of designing and building small power stations. They also began to inspect and then provide management services for utility companies that had fallen into the General Electric orbit. This eventually led to building and financing new power companies, as well as undertaking a great variety of other industrial projects—power plants, steel mills, water works, and even the new MIT campus when that institution relocated from Boston to Cambridge. By the 1930s Stone & Webster was a major enterprise that employed nearly a thousand engineers and draftsmen. During WW II it built synthetic rubber and alcohol plants, explosive factories, and the nuclear facilities at Hanford, Washington and Oak Ridge, Tennessee. It continues today as a major international constructor of pipelines, coal mines, oil refineries, and other advanced industrial plants.[2]

This pattern is by no means extinct today. Many firms are still started by engineers who have parlayed their technical skills into a thriving business. The multitude of electronics plants that have been created by electronics engineers (many of them former employees of large corporations) have received the most interest, but examples can be found in many other high-tech sectors, and some low-tech ones as well.

Modern engineer-proprietors often experience the feeling of being pulled between engineering and managerial roles. Many have an acute awareness that their identity as practicing engineers is being compromised when they spend most of their working hours with lawyers and accountants rather than technical personnel. A hundred years ago the duality of roles was less problematic. Serving as both manager and engineer simply followed from being the owner of a firm. Tasks were less specialized, and the concept of management as a distinct career with its own techniques and way of looking at the world had not yet appeared. This began to happen only towards the end of the nineteenth century as a consequence of the growing size and complexity of industrial firms. These changes called for more sophisticated techniques of management. At the same time, the ownership of these firms was becoming more diffuse; a growing number of firms were legally owned by a multitude of stockholders rather than a single entrepreneur, making it difficult for many managers to use ownership as the basis of their authority. Management was thus faced with the need to develop a whole ensemble of managerial techniques while at the same time finding a justification for their use. Strengthening the connection between engineering and management seemed to hold the solution for both of these problems. As new cadres of management attempted to deal with the myriad problems of administering large and complex enterprises, some key aspects of management began to resemble technical activities rooted in supposedly scientific principles. In other words, management could be seen as a kind of engineering. As such, it became technically more sophisticated while, equally important, it gained legitimacy through its presumed connections with science.

The ideology that looked upon management as an extension of engineering was not just based on theoretical speculation; many of the early managers who were imbued with this spirit were in fact drawn from the ranks of practicing engineers.

This can be clearly seen in one of the pre-eminent modern industries of the nineteenth and early twentieth century industries, the railroad. A list of presidents of mid-19th century railroads shows a very strong representation of men who had first trained and worked as civil engineers: in Chapter Seven we met Moncure Robinson (Richmond, Fredricksburg, and Potomac), John B. Jervis (Michigan Southern), Herman Haupt (Pennsylvania), George W. Whistler (Western), and Benjamin Latrobe (Baltimore and Ohio). To their names we could add David C. McCallum (Erie), J. Edgar Thomson (Pennsylvania), and George B. McClellan (Illinois Central). The premier "high-tech" enterprise of its era, the railroad was an excellent proving ground for the techniques of management that had been developed by men with engineering training and experience.

The merger of engineering and management spread into many other industries. The American steel industry owed much of its early development to the efforts of William Holley (1832-1882). An engineering graduate of Brown University, Holley introduced the Bessemer process to the United States, at the same time arranging a patent pool with William Kelly, an American who had invented an almost identical process. Holley's managerial innovations were no less important. He gave much thought to the spatial arrangements of a steel mill and the routing of materials within it. The result was a considerable gain in productivity when compared with contemporary American and British steel mills. Holley was convinced that superior technical arrangements were inseparable from advanced managerial methods: "Better organization and more readiness, vigilance and technical knowledge on the part of the management have been required to run works up to their capacity, as their capacity has become increased by better arrangement and appliances."[3]

The swiftly developing chemical industry provided another important example of engineers playing a vital role in management. The most notable was Dupont, where engineers (many of them graduates of the Massachusetts Institute of Technology) were the driving force behind the firm's expansion and organizational development.[4] The chemical industry in general proved to be a fertile field for the fusion of engineering and management. Chemical engineering as a distinct occupation was from the outset predicated on the engineers' combination of technical and managerial roles.[5] In seeking to distance themselves from chemists, early chemical engineers stressed that managerial skills and responsibilities separated them from the former. As the president of the American Institute of Chemical Engineers claimed in 1920, a chemical engineer was "a chemical manager, that is, one who manages enterprises requiring the knowledge and application of chemistry." For successful chemical engineers, mere technical knowledge was not enough; effective management of workers was an essential part of their role. As one chemical engineer crudely put it, "Of what consequence is the chemical reaction which depends on labor to make it work, if the chemist does not know how to make a 'Dago' work?"[6]

What made the contributions of this new breed of engineer-manager important was not simply the specific technical skills that they brought to their jobs, but their championing of rationalizing spirit that sought to put management on a firm "scientific" basis. A crucial component of this spirit and the managerial practices that grew out of it was an abiding concern with cost accounting as an essential tool of management. As we have seen in the previous chapter, engineering rarely takes place in a financial vacuum. Nor do most engineers resent their submission to financial discipline, for a concern with budgetary matters has always been an integral part of engineering practice; one definition of an engineer is "a person who can design something to be made for a quarter that any fool can design to be made for fifty cents."

As managerial concerns became increasingly prominent among engineers, financial subjects were tightly woven into engineering training and practice. The methodology of precise cost accounting was first elaborated in journals specifically oriented to engineers, most notably the *Transactions* of the American Society of Mechanical Engineers, *American Machinist*, and *Engineering News*.[7] Significantly, the latter changed its name to *Industrial Management* in 1916.[8] *The Engineering Magazine* was a relatively insignificant publication until in 1896 it struck what its publisher characterized as the "golden vein": a series of articles by Horace Arnold on "Modern Machine Shop Economics." This journal subsequently became a key source for the dissemination of information on industrial management.[9]

During the late nineteenth and early twentieth centuries, growing numbers of engineers were recruited into management. After examining the career patterns on engineers from 1884 to 1924, the influential Wickenden report on engineering education noted with pride "a healthy progression through technical work towards the responsibilities of management."[10] From 1900 to 1930, between two-thirds and three-quarters of engineering graduates had taken on managerial positions within fifteen years of their graduation.[11]

The Massachusetts Institute of Technology was a particularly important source of top-level managers with engineering training. Its purpose was made clear at its foundation; it was conceived as an institution "intended for those who seek administrative positions in business... where a systematic study of political and social relations and familiarity with scientific methods and processes are alike essential."[12] During the first half of the 20th century, half of MIT's engineering graduates went into business and management careers.[13] One particularly impressive cohort of MIT graduates boasted the chief executives of General Motors, General Electric, Dupont, and Goodyear Tire and Rubber.[14]

Engineering thus became one of the major influences on the beliefs and practices characteristic of the emerging profession of management. At the same time, for the most ambitious engineers more was at stake than simply the administration of individual firms; engineers had no less a role than guiding and organizing all aspects of the industrial economy. In fulfilling this obligation, the engineer had the duty of ensuring that "the bearings of our human industrial structure are properly designed, properly constructed, and properly adjusted."[15]

In the United States, schools of engineering did much to reinforce this trend by making management an integral component of engineering training. By 1932, 35 engineering schools were providing formal instruction in management subjects, including many topics far from the technical concerns of engineering: merchandising, finance, office administration, cost accounting, and personnel.[16] All in all, it is only a slight exaggeration to claim that "modern management was the product of engineers functioning as managers."[17]

Engineers as Managers: The Soviet Union

While engineering came into close alignment with management under capitalist aegis in Europe and North America, the same thing was happening in the radically different social and economic system of the Soviet Union. Despite civil war, foreign invasion, and Stalinist terror, Soviet managerial principles and practices were deeply influenced by engineering training and work. More than anywhere else, management in the Soviet Union was the domain of people with an engineering background. Most Soviet enterprise managers were trained as engineers. Even more significantly, so thoroughgoing was the penetration of engineering into top-level management that by 1966 sixty-five percent of the members of the Communist Party's Central Committee had a post-secondary technical education, and most of them had experience as engineers and industrial managers.[18] In the higher-level Politburo, no fewer than 80 percent of the membership had this background in the 1960s.

On the face of it, this is a curious phenomenon, for the egalitarian component of Marxist ideology hardly seems favorable to the establishment of a technical elite whose expertise could be a new source of domination and inequality. But in fact the Soviet leadership was always far more concerned with the rapid development of an industrial economy than with the development of an egalitarian society. Under these circumstances the latter was even seen as an obstacle to the former, and "equality mongering" was attacked as actually being counter-revolutionary.

As a result of the all-out commitment to industrialization and technological modernization, engineers were favored members of Soviet society. At first they were fairly well insulated from political pressures; during the New Economic Policy period (1921-26) most engineers were entrusted with the oversight of day-to-day technical matters, while the managers of their enterprises concerned themselves with overall supervision.[19] This began to change in the late 1920s when, with Stalin firmly in control, many engineers experienced harassment and persecution. Still, the assault on the technical specialists did not represent an attack on an incipient technocracy; rather, it was directed against older engineers who were viewed as the remnants of the old elite of Tsarist days. Many of them suffered purges and harassment,[20] but a sizeable portion of the old technical specialists continued to play leading roles in industry and education. Even so, it was clear that engineers could no longer immerse themselves in technical matters and remain insulated from political demands. Stalin made this quite clear at an address before a managers' conference in early 1931:[21]

> Ten years ago we had the following slogan, 'Since communists don't yet properly understand the technique of production, since they have still to learn how to administer the old economy, let the old technicians and engineers... manage production, and you, communists, don't interfere in technical matters...' Now it is time to finish with the stale view that we should not interfere in production. It is time we mastered another, new idea corresponding to the present period: *interfere in everything*.

With the passage of time the class position of engineers changed as a new cadre of engineers emerged from the ranks of industrial workers. These new engineers were products of the new communist social order, and many of them combined political activism with technical expertise. But political activism had its hazards. When the great purges of the late 1930s decimated the political and managerial elite, it was the 'red' technical specialists who suffered most. Many of them were demoted, or in many cases executed or exiled, their places taken by engineers who had not been politically active.[22]

A politically neutralized cadre of engineers had become a bulwark of the Soviet economic order. An engineering education was one of the main paths to a career in management, although social background and party membership continued to be of considerable importance.[23] Drawn from a socially acceptable class, and having no ties to the either the pre-Bolshevik elite or the leadership deposed by Stalin, this new group of engineers and other technical experts occupied a key position in Communist society. Working in close conjunction with the political leadership they enjoyed considerable power and influence, but at the same time they were always subject to the authority—and the interference—of the political leadership.

The massive efforts at achieving rapid economic growth through centralized economic planning presented great opportunities as well as great dangers to engineers involved in industrial management. Management was anything but a pure technical exercise; it often entailed political decisions of a high order. The Communist Party was not about to give up its economic leadership, and especially in the Stalinist era it made abundant use of terror in order to maintain its control over economic planning and management.[24] This situation produced many problems for Soviet engineers. Although engineering offered one of the surest routes of upward mobility, it was a difficult and hazardous occupation, for engineers could never be certain of their authority or even their personal safety

For all the stresses they endured, Soviet engineers enjoyed high status in Soviet society. Many of them belonged to the political elite through their membership in the Communist Party. But political affiliation aside, engineers and other technical experts occupied a pivotal position in Soviet society because many of their key goals and values were congruent with those of the leadership. The Communist ruling elite had a strong commitment to economic and technical modernization, yet at the inception of the Soviet Union it had no clear program for achieving these goals, nor did it have any experience in organizing a rapidly industrializing society. These tasks fell to the engineers and other members of the technical intelligentsia, both the old specialists and the new breed of Soviet engineer.

The vision and the technical expertise of engineers were of immense importance in making the Soviet Union a formidable—if fatally flawed—industrial and military power.

In this milieu, engineers enjoyed many privileges and a high level of respect. Even so, engineers as a group were not part of the ruling elite simply by virtue of their technical abilities and accomplishments. Only a few of them gained entrance into the upper ranks of the political leadership, and those that did were not necessarily the most technically proficient. As Kendall Bailes' study of the Soviet political elite of the 1960s indicates, engineers and other members of the technical intelligentsia who did achieve political pre-eminence differed in a number of significant ways from those who had distinguished themselves solely in the technical realm. A technical background came close to being a necessary condition for gaining political authority, but it was by no means a sufficient one. The achievement of a high political position was closely correlated with worker or peasant class origin. Moreover, the political elite was comprised of males drawn largely from those of Slavic stock, while ethnic minorities and women were greatly underrepresented. Its members were also far more likely to have received their technical education at provincial universities. Finally, as might be expected, their occupational roles usually centered on direct production work, in contrast to the academic and R&D roles of the non-elite group.[25]

Although technical expertise was of vital importance, the Soviet Union was never a technocracy. A great deal of management, both political and economic, was the work of people with an engineering background, but their skills and accomplishments as engineers did not easily lead them into positions of authority. For Soviet engineers the translation of knowledge into power was always problematic.

Scientific Management

In its search for the optimal methods of promoting rapid economic and technological modernization, the Soviet leadership eagerly seized on Frederick Taylor's system of Scientific Management. Its use was thought to be of crucial importance, for as Lenin saw it, "The possibility of socialism will be determined precisely by our success in combining Soviet government and the Soviet organization of administration with the modern achievements of capitalism. We must organize in Russia the study and teaching of the Taylor System and systematically try it out and adopt it to our purposes."[26]

Scientific Management proved to be of much less importance than Lenin had anticipated, and even in the capitalist world it never lived up to the grandiose expectations of its founder and his followers. Even so, Scientific Management is an important topic for the history of engineering, for it represented the most comprehensive effort to make all of management a branch of engineering. For the engineers who created and attempted to apply it, Scientific Management was more than a technique of factory organization; it offered the possibility of managing the whole of society. According to Taylor, his system "can be applied with equal force to all social activities: to the management of our homes; the management of our farms; the management of the business of our tradesmen large and small; of our churches, of our philanthropic organization, our universities; and our government departments."[27]

The emergence and spread of Scientific Management bear witness to its origins in engineering. Before he devoted himself to the elaboration and promulgation of his system, Frederick W. Taylor (1856-1917) had achieved notable successes while serving as chief engineer of the Midvale Steel Company in Pennsylvania. There he conducted extensive research on the technology of metal working that led to the development of high-speed tool steel and optimal techniques for cutting metal. In recognition of his accomplishments, Taylor was elected president of the American Society of Mechanical Engineers. But Taylor's ambitions far surpassed even these noteworthy achievements. He was convinced that just as scientific principles could be used to control the shaping of obdurate metals, so too could they be used to control an even more recalcitrant resource: labor. Through the use of carefully timed measurements of workers' activities and the planning of work by expert industrial engineers, efficiency and output would undergo a vast expansion. Unlike traditional capitalist practices, increased profits would not come through holding down wages. To the contrary, the greater efficiency brought by Scientific Management would result in increased production, thereby allowing a larger surplus to be shared by workers and capitalists. Under these circumstances, class conflict would become irrelevant. In Taylor's words:[28]

> The great revolution that takes place in the mental attitude of the two parties under scientific management is that both sides take their eyes off the division of the surplus as the all-important matter, and together turn their attention to increasing the size of the surplus until this surplus becomes so large... that there is ample room for a large increase in wages for the workman and an equally large increase in profits for the manufacturer.

Although class conflict supposedly would fade away under Taylor's system, new divisions would appear. Scientific Management was predicated on the radical separation of thinking and doing, of planning and execution. As Taylor explained his system:[29]

> The work of every workman is fully planned out by the management at least one day in advance, and each man receives in most cases complete written instructions, describing in detail the task which he is to accomplish, as well as the means to be used in doing the work. And the work planned in advance in this way constitutes a task which is to be solved... not by the workman alone, but in almost all cases by the joint effort of the workman and management. This task specifies not only what is to be done, but how it is to be done and the exact time allowed for doing it.

Efficient work required the exercise of control by management over many things that hitherto had been the province of the worker. Very little was to be delegated to the workers; even the skills they had amassed had to be systematized and administered by the manager: "The managers assume... the burden of gathering of all of the traditional knowledge which in the past has been possessed by the workman and then of classifying, tabulating, and reducing this knowledge to rules, laws, and formulae..."[30]

The centrality of engineers to the working of this system should be evident. When the Taylor Society was established to expound the principles of Scientific Management, its membership was largely restricted to engineers.[31] Scientific Management required a virtuous cadre of disinterested technical experts. According to Taylor's doctrine, the engineers who implemented Scientific Management served neither labor nor capital; it was their task to maintain the rule of science in the factory, defending it against "narrow vision and vested interests" of workers and employers alike.[32] Taylor's system, in addition to eliminating labor strife and bringing vastly higher levels of production, would install engineers as the directors of modern society. As Henry L. Gantt, one of Taylor's early disciples, noted, "the engineer, who is a man of few opinions and many facts and many deeds, should be accorded the economic leadership which is his proper place in our economic system."[33]

This expectation remained unrealized. Although it is highly doubtful that Taylor's principles would have brought the benefits claimed for them, Scientific Management was not fully tested, for it never was fully put into practice in any organization. As might be expected, Taylor's system ran into considerable opposition from the workers. But no less important, Scientific Management was resisted by management itself. It had some influence over managerial theory and practice, but only when the total system was diluted into a set of principles that were selectively applied by management.[34] The complete system of Scientific Management as envisaged by Taylor and his disciples could not be put to use, for it directly conflicted with the interests of entrenched owners and managers, who were not about to abdicate their authority to mere technical expertise.[35] As Taylor noted at the first Scientific Management conference, "...in the work of changing from the old to the new system, nine-tenths of our troubles are concerned with those on the management side, and only one tenth with the workers. Those in management are infinitely more stubborn, infinitely harder to make change their ways than are the workers."[36]

Yet while as Scientific Management was ebbing, even more grandiose notions of engineers taking over human affairs were being formulated, and for a short while thereafter the political ideology of Technocracy emerged as a bold attempt to make engineering the center of economic and political power.

Technocracy

Scientific Management was part of a larger intellectual movement that emerged in the early twentieth century as engineering methods were put forth as the best means of solving problems and organizing people. As one electrical engineer put it in 1917, "It matters not whether the problems before him are political, sociological, industrial or technical, I believe that an engineering type of mind is best fitted to undertake them."[37]

Although the technocratic spirit became fully manifest in the twentieth century, its historical roots go considerably deeper. In one sense, it is a variant of Plato's idea of rule by philosopher-kings. It appears in a more concrete form with the French political theorist Henri Comte de Saint-Simon (1760-1825), whose ideas have been dubbed "the religion of engineers."[38] Saint-Simon championed the emergence of a new society in which the old political-religious elite would be supplanted by scientists and industrialists. Their knowledge and application of physical and social science would be used to organize society and to create its moral foundation. To provide an institutional basis for the work of the new elite, Saint-Simon proposed the establishment of an "industrial parliament" where scientists, engineers, and industrialists would plan and supervise the public projects that would be the foundation of the economy.[39]

Technocratic theory gained new momentum during the early 20th century through the writings of the American economist-sociologist Thorstein Veblen (1857-1929).[40] As Veblen analyzed modern society, the basic source of conflict in modern societies derived from the tensions inherent in an industrial system driven on the one hand by rationality and the search for efficiency, and on the other by a business system that was exclusively concerned with the amassing of profit. The former was the domain of the engineer, who embodied "the instinct of workmanship," while the latter was the realm of a leisured class of businessmen whose prime interests lay in all forms of conspicuous consumption. The implications of Veblen's analysis were quite clear: the prevailing system of ownership and management was an obstacle to social and economic progress. The potential of modern technology to make a better world could only be realized when engineers occupied their rightful place as the chief managers of society. To this end, Veblen proposed a "Soviet of Technicians," a vaguely defined system of administration that would come to power through either a general strike of workers and technicians, or the simple abdication of the business elite.[41]

While Veblen presented his abstract (and some might say naive) formulations, others attempted to convert his ideas into reality. One of Veblen's associates, an engineer of dubious qualifications named Howard Scott, formed an association known as the Technical Alliance in 1919. This organization recruited a impressive list of participants, most notably Charles Steinmetz (1865-1923), the legendary mathematician and engineer who did so much to advance electrification in the late nineteenth and early twentieth centuries. The Alliance's activities centered on the conduct of an empirical study into the extent of waste in industry, an assessment of the resources necessary to sustain society at given levels of comfort, and a graphic rendition of how the present economic system operated. The ultimate goal of these efforts was to generate data that would provide a foundation for technocratic governance.[42] This extremely ambitious project, which intended to survey no fewer than 3000 industries, never get off the ground, and in any event the prosperity of the 1920s seemed to render the issues moot.

The Depression of the 1930s created a radically different climate, one apparently ripe for new forms of political and economic management. For a brief period an organized Technocracy movement headquartered at Columbia University's Department of Industrial Engineering and directed by Scott received widespread attention. Yet after this brief flurry of excitement the movement collapsed. The proximate cause was Scott's incompetent leadership, but more fundamentally, the Technocracy movement failed because it was rent by internal disputes, the inability to formulate a coherent plan of action, and a lack of interest in organizing a mass movement.[43]

Although it made the rule of the engineer the centerpiece of its program, the Technocracy movement was animated by more general beliefs and ambitions. Much of its appeal could be traced to the Progressive Era's earlier attempt to substitute detached administration for sordid politics, coupled with a faith in science and technology as a panacea for all economic and social ills.[44] The movement also hoped to draw upon the energies and enthusiasms of the rapidly growing engineering profession. Along with an increase in the number of engineers had come an expansion in influence, leading to on occasion to hyperbolic statements about the emerging role of the profession: "The engineers more than all other men, will guide humanity forward until we have come to some other period... On the engineers and on those who are making engineers rests a responsibility such as men have never before been called on to face."[45]

In fact, engineers had come nowhere close to realizing these lofty ambitions, for the growth of the profession had not been matched by increased political influence and social status. Most engineers did not seem to be especially attracted to the political destiny that the Technocrats had prepared for them. Technocracy offered engineers a chance to vault to the highest echelons of political rule, yet even during its brief heyday in the early 1930s the Technocracy movement counted very few engineers in its membership. Its backbone of support was drawn from other inhabitants of the middle class: doctors, lawyers, middle managers, and a few small businessmen. Despite strenuous efforts to recruit them, engineers were conspicuous by their absence.[46]

Some engineers stayed away because the Technocracy movement with its stress on the disruptive aspects of technological change (most notably technological unemployment) seemed to threaten the self-image of engineers as agents of progress. Thus, in rejecting Technocracy, one professional body, the Council of Engineers, made the remarkable statement that "nothing inherent in technological improvement entails economic and social maladjustments."[47] Such a belief might be expected given the Council's strong pro-business orientation, but even rank-and-file engineers demonstrated their basic conservatism by rejecting a movement that seemingly catered to their interests. The fact that Technocracy proved to be of limited appeal to engineers seems to indicate that their ambitions lay elsewhere. Many engineers aspired to positions of greater status and influence, but as they saw it, these would not come through quixotic attempts to seize political power, but from working within the existing system. Engineers extended their influence by assuming managerial roles within the firms they served and not through attempting to grab the reins of government.

Engineers and Modern Management

The trends that emerged during the early decades of the 20th century continue to this day as large numbers of engineers are drawn into management. Even when they don't occupy formal managerial positions, the majority of engineers have administrative duties in addition to their technical ones. One study of engineers in six industrialized nations found that the percentage of engineers who were involved in personnel selection ranged from 40 to 69 percent. For planning and controlling budgets, the figures were 59 and 77 respectively, while between 48 and 83 percent were involved in directly supervising others.[48]

For some engineers, taking on managerial responsibilities is undoubtedly an annoyance, but for the majority it seems to be a natural stage in an engineering career. For many, engineering is only a way station on the road to a career in management. Throughout the industrial world, 15 to 30 percent of engineers have management and administration as their major responsibility.[49] This pattern has been especially pronounced in Continental Europe where as many top executives have been drawn from the ranks of engineering and science graduates as from those of business and economics.[50]

Rather surprisingly, Japan, whose economy is often viewed as being driven by technical expertise, presents a contrary picture. Some scholarly accounts have indicated a close connection between an engineering education and a career in upper management,[51] but this is not borne out by reliable statistics. One Japanese study conducted in the early 1970s found that only 23 percent of chief executives had either an engineering or science background. Similarly, a 1981 Ministry of Education study indicated that only 4.5 percent of new engineering graduates went directly into managerial or clerical positions. Engineering does not seem to be a favored vehicle for attaining a managerial position in Japan; indeed, the proportion of Japanese chief executives with a background in engineering may be the lower than all Western industrial societies.[52]

It is of course possible and even likely that many Japanese engineers do move into at least the middle levels of management. It is certainly the case that in the United States and Europe the career of the engineer seems to logically progress towards management. On both continents half of the engineers who have attained the age of 55 occupy managerial positions. Research, development, design, and testing — the essence of engineering — tend to be the province of younger engineers.[53] It has even been argued that to remain in engineering is to have a career that is at a dead end; a number of years ago an occupational sociologist asserted that "the engineer who, at forty, can still use a slide rule or a logrithmic table, and make a true drawing, is a failure."[54] This is surely an overstatement. Still, it is likely that the reward structures of most organizations do not serve the needs of technically oriented employees who are uninterested in management positions.[55]

Is the passage from engineering into management the result of dissatisfaction with engineering work and its rewards, or is it something that engineers build into their career expectations from the beginning? Human motivation is of course a tricky thing to dissect. People enter engineering for many reasons, not all of them consciously apparent even to themselves. Still, it seems to be the case that more than other knowledge-based occupations, engineering attracts people with a strong interest in social and economic advancement. In comparison with scientists, for example, engineers are more concerned with opportunities for high earnings, career advancement, the exercise of leadership, and the possibility of making a solid contribution to the the success of their organization.[56] It also seems to be the case that engineers select their employers almost as much for their advancement potentials as for the interesting work they offer.

More concretely, many people go into engineering because it affords an opportunity to eventually join the ranks of management.[57] A management position offers financial rewards and status that are out of the reach of all but a few engineers. But money and status are not the only motivators. The nature of many engineering tasks may also impel many engineers to aspire to a managerial position. Engineers are often people with a high need for achievement, yet a great deal of engineering work can be routine and trivial. Consequently, "It seems likely... that many engineers climb the managerial ladder not necessarily because they are more interested in people than in things, but because their creative energies are not being fully utilized in engineering work."[58] In similar fashion, many technically proficient people strive for managerial positions because the autonomy these positions confer makes it easier to exercise technical competence.[59] Accordingly, many engineers enter the ranks of management so that they can be more effective *as engineers*. As one engineer put it, moving into management "means that your individual viewpoint on what should be done becomes more important and you get a better chance to use it."[60] A managerial position confers "clout," giving an employee the opportunity to see his or her technical abilities translated into something substantial. It also provides a more inclusive organizational role, one that permits involvement in a wider array of technical activities and accomplishments.

Not all engineers have seen in the assumption of managerial roles the surest means of achieving influence and high status. There is an alternative possibility: securing for engineers the exalted position of the independent professional. Yet this has proved to be a quest that in many ways has been just as quixotic as the attempt to become the governors of a technocratic social order. This still-unfulfilled search for professional status will be explored in the last chapter. But first we need to consider an essential element of professionalization: participation in a specialized educational program. The history of engineering education will accordingly be the topic of the next chapter.

Notes

1 Daniel H. Calhoun, *The American Civil Engineer: Origins and Conflict* (Cambridge, Massachusetts: MIT Press, 1960), p. 77. During this period a distinction could be made between "native" engineers, who served as proprietor-manager and "foreign" engineers, who simply sold their services and did not have a financial stake in the firm for which they worked. (Ibid., p. 23)

2 Sam Bass Warner, Jr., *Province of Reason* (Cambridge: Harvard University Press, 1984), pp. 52-66.

3 Alfred D. Chandler, Jr., *The Visible Hand: The Managerial Revolution in American Business* (Cambridge, Massachusetts: Harvard University Press, 1977), p. 260.

4 Ibid., p. 4.

5 Terry S. Reynolds, "Defining Professional Boundaries: Chemical Engineering in the Early Twentieth Century," *Technology and Culture* 27, 4 (October 1986), pp. 712-16.

6 Ibid., p. 714.

7 Chandler, op. cit., pp. 464-65.

8 Ibid., p. 466.

9 Daniel Nelson, *Managers and Workers: Origins of the New Factory System in the United States 1880-1920* (Madison: University of Wisconsin Press, 1975), p. 49.

10 David Noble, *America by Design* (New York: Knopf, 1977), p. 41.

11 Ibid., p. 310.

12 Eric Ashby, *Technology and the Academics: An Essay on Universities and the Scientific Revolution* (Oxford: The Clarendon Press, 1982), p. 61.

13 John B. Rae, "Engineering Education as Preparation for Management: A Study of MIT Alumni," *Business History Review*, 29, 1 (March 1955), p. 61.

14 Paul W. Litchfield, *Industrial Voyage* (Garden City, N.Y.: Doubleday, 1954), p. 55.

15　Noble, op. cit., p. 312.

16　Ibid., p. 313.

17　Ibid., p. 263.

18　Kendall E. Bailes, *Technology and Society under Lenin and Stalin* (Princeton: Princeton University Press, 1978), pp. 6 and 267-68.

19　Nicholas Lampert, *The Technical Intelligentsia and the Soviet State: A Study of Soviet Managers and Technicians, 1928-1935* (London: Macmillan, 1979), p. 30.

20　Steve Smith, "Taylorism Rules, OK? Bolshevism, Taylorism, and the Technical Intelligentsia in the Soviet Union, 1917-1941," *Radical Science Journal*, 13 (1983), pp. 5-8.

21　Lampert, op. cit., p. 85.

22　Lampert, op. cit., pp. 152-53.

23　Bailes, op. cit., p. 307.

24　Ibid., p. 270.

25　Ibid., pp. 412 and 431-41.

26　V.I. Lenin, "The Immediate Tasks of the Soviet Government," *Izvestia*, 28 April 1918, in V.I. Lenin, *Selected Works*, vol. 2 (Moscow: Foreign Languages Publishing House, 1947), p. 327.

27　Quoted in Samuel Florman, *The Existential Pleasures of Engineering* (New York: St. Martin's, 1976), p. 8.

28　F.W. Taylor, *Scientific Management, Comprising Shop Management, the Principles of Scientific Management, Testimony before the Special House Committee* (New York, 1911), quoted in Charles S. Maier, "Between Taylorism and Technocracy: European Ideologies and the Vision of Industrial Productivity in the 1920s," *Contemporary History* 5, 2 (1970), p. 32.

29　Frederick W. Taylor, *The Principles of Scientific Management* (New York, W.W. Norton, 1967. Originally published in 1911), p. 39.

30　Ibid., p. 36.

31　Samuel Haber, *Efficiency and Uplift: Scientific Management in the Progressive Era, 1890-1920*. Chicago and London: University of Chicago Press, 1964), p. 32.

32　Ibid., p. 28.

33　Ibid., p. 32.

34 Peter Meiksins, "The 'Revolt of the Engineers' Reconsidered," *Technology and Culture* 29, 2, (April 1988), p. 242.

35 Nelson, op. cit., p. 75.

36 Donald R. Stabile, *Prophets of Order: The Rise of the New Class, Technology, and Socialism in America* (Boston: South End Press, 1984), p. 47.

37 Quoted in Edwin T. Layton, Jr., *The Revolt of the Engineers: Social Responsibility and the American Engineering Profession* (Baltimore and London: Johns Hopkins University Press, 1971), p. 66.

38 The phrase appears in Friedrich Hayek, *The Counter Revolution of Science* (Glencoe, The Free Press, 1955), and is quoted in Ian A. Glover and Michael P. Kelly, *Engineers in Britain: A Sociological Study of the Engineering Dimension* (London: Unwin Hyman, 1987), p. 13.

39 Irving M. Zeitlin, *Ideology and the Development of Sociological Theory* (Englewood Cliffs, New Jersey: Prentice-Hall, 1968), pp. 56-69.

40 Thorstein Veblen, *Engineers and the Price System* (New York: Huebsch, 1921)

41 William E. Akin, *Technocracy and the American Dream: The Technocrat Movement, 1900-1941* (Berkeley, Los Angeles, and London: University of California Press, 1977), pp. 24-25.

42 Ibid., p. 34.

43 Ibid., pp. 80-115.

44 Haber, op. cit.

45 Akin, op. cit., p. 8.

46 Ibid., p. 107.

47 Ibid., p. 159.

48 Wouter Van den Berghe, *Engineering Manpower: A Comparative Study of the Employment of Graduate Engineers in the Western World* (Paris: UNESCO, 1986), p. 171.

49 Ibid., p. 142.

50 David Granick, *The European Executive* (Garden City, New York: Doubleday, 1972)

51 See, for example, G.C. Allen, *The Japanese Economy* (New York: St. Martin's, 1981), p. 95, in which it is noted that engineers occupied two-thirds of the seats on the boards of directors of Japanese companies. But this is not a meaningful figure, for it refers not to practicing and

formerly practicing engineers, but simply to those who have received an education in some area of applied science.

52 Earl H. Kinmonth, "Engineering Education and Its Rewards in the United States and Japan," *Comparative Education Review*, 30 (August 1986), pp. 400-401.

53 Joel E. Gerstl and Stanley P. Hutton, *The Anatomy of a Profession: A Study of Mechanical Engineers in Britain* (London: Tavistock, 1966), p. 91; Van den Berghe, op. cit., pp. 181-82.

54 Everett C. Hughes, *Men and Their Work* (Glencoe: The Free Press, 1958), p. 137. There is some evidence that the most technically competent engineers remaining in lower-level positions (non-managerial ones in most cases) have the greatest degree of job dissatisfaction found among engineers. See Lotte Bailyn, *Living with Technology: Issues at Midcareer* (Cambridge, Massachusetts: MIT Press, 1980), p. 52.

55 Bailyn, op. cit.

56 Ibid., p. 25.

57 Robert Perrucci and Joel Gerstl, *Profession without Community: Engineers in American Society* (New York: Random House, 1969), p. 140.

58 Gerstl and Hutton, op. cit., p. 124.

59 Bailyn, op. cit., p. 54

60 R. Richard Ritti, *The Engineer in the Industrial Corporation* (New York and London: Columbia University Press, 1971), p. 25.

Chapter 10

Educating Engineers

For much of its history, engineering was learned on the job. As we have seen, most of the great engineers of the past received little in the way of formal education. Their expertise was the product of innate talents and the opportunity to use and extend these talents while engaged in actual engineering work. As with many other aspects of engineering, this began to change dramatically towards the end of the nineteenth century, as more and more engineers received formal, systematic training before embarking on their careers. Part of the reason for this lay in the greater complexity of engineering and its closer association with science. Experience and rule-of-thumb methods could be quite effective in the design of water wheels or textile machinery, but during the nineteenth century the growing sophistication of machinery, coupled with the emergence of the chemical and electrical industries, changed the nature of engineering by requiring a greater familiarity with scientific principles. Still, the need for scientific knowledge was not the only thing impelling a more structured approach to the training of engineers. As we shall see, the new patterns of engineering education were closely tied to the social and organizational changes that were altering every facet of engineering.

The Development of Engineering Education in Europe

As was noted in Chapter 5, the first systematic effort at formally training engineers occurred in France with founding of the *Ecole Nationale des Ponts et Chaussees* in 1747. The French government also attempted to extend its military capabilities through the establishment of the *Ecole du Corps Royal du Genie* in 1749. This institution offered advanced training for military engineers through a rigorous curriculum that included cartography, building construction, machine design, and the construction of fortifications. It also boasted the best-equipped chemistry and physics laboratories in all of Europe.[1] It was at this school that the great mathematician Gaspard Monge developed his system of teaching descriptive geometry. His graphic methods allowed one to use a ruler and compass to solve a varied set of practical problems, such as the rendering of solids in two dimensions, the shaping of vaults, and the intersection of surfaces.[2] Monge's methods remain the foundation of mechanical drawing, long a staple of engineering education and practice.

An even greater educational achievement came with the French Revolution, which was accompanied by a considerable enthusiasm for using science and technology to perfect humanity. Inspired by Monge and Lamblardie (at the time the head of the *Ecole des Ponts et Chaussees*), a comprehensive school of engineering was founded in 1794 as the *Ecole des Travaux Publics*. A year later it was renamed the *Ecole Polytechnique*.[3] It soon boasted the best scientific faculty anywhere in the world and a superb student body that had been selected through rigorous entrance examinations. In the forty years following its founding, more than 14,000 candidates took entrance examinations to the *Ecole*; only 5,502 were accepted.[4] Its graduates went on to occupy some of the highest government positions; indeed, many top governmental positions were virtually monopolized by *Ecole Polytechnic* graduates. Many of them went on to more specialized engineering training at advanced *ecoles d'application* that offered specialized training in such fields as mining, bridge design, and artillery.

While the *Ecole Polytechnique* provided a rigorous training for some of the best minds in France, it was often criticized for its excessively abstract and theoretically oriented curriculum. For the critics, the education imparted by the *Ecole Polytechnique* was too rarified to be of much use for engineers working in manufacturing industries. But it was not the only institution providing an engineering education. A less elite-oriented education was provided by various *ecoles d'arts et metiers*. Founded during the reign of Napoleon to train young men of modest means to be skilled workers and low-level supervisors, these schools eventually specialized in practical engineering training. By the 1860s about 5000 of their graduates—40 percent of France's trained engineers and middle-level technicians—were working in French industry, business, and transport.[5] Many of them played important roles in management as well as engineering, and most had attained managerial and supervisory positions before turning 30.[6]

Private initiatives were also important in extending French engineering education. The restored Bourbon monarchy that had come to power after Napoleon's downfall was hostile to the emerging class of industrialists, and it had little interest in promoting education in general. Some of the void was filled by a group of businessmen when they founded *The Ecole Centrale des Arts et Manufactures* in 1829 (the school was ultimately taken over by the French state in 1857). Because it was oriented more toward the training of working engineers, the course of study at the *Ecole Centrale* stressed the application of science and mathematics in the laboratory and the shop.[7] Soon after its inception the *Ecole* became the largest source of civil engineers in France, a distinction it holds to this day.[8]

Engineering education was quite different in Great Britain, for the role of educational institutions was far less prominent. Although engineering topics had been presented in British universities under the rubric of Natural Philosophy during the late eighteenth century, many decades passed before engineering education was put on a firm institutional footing. The foundation of University College, London (1826) and King's College, London (1828) marked an important turning point, for these institutions emphasized science and engineering from the beginning. By 1838, King's had a department of engineering, while University College established a chair in engineering a few years later.[9] In contrast, Oxford and Cambridge maintained their commitment to a classically based education through much of the nineteenth century. Courses of study approximating engineering (statics and dynamics, pneumatics, optics, hydrostatics, and polarization) were given by G.B. Airy, the Lucasian Professor of Mathematics at Cambridge in the 1830s, but the first true engineering position at Cambridge, the Chair of Applied Mechanism and Applied Mechanics, was not established until 1875. Oxford did not get around to establishing a chair in engineering until 1907.[10]

While English engineering education lagged, a somewhat more vigorous attempt was being made in Scotland. Technical training on a fairly sophisticated level was done at schools such as the Glasgow Mechanics Institution and Anderson's Institution (sometimes known as Anderson's College or, more grandly, the Andersonian University). The latter was created from a small bequest of Professor John Anderson, whose tenure at the the University of Glasgow had been marked by many academic and personal disputes. Still, these institutions lay outside the orbit of traditional Scottish Universities. An acceptance of engineering as a legitimate field of academic study came with the creation of a Chair in Engineering at the Univeristy of Glasgow in 1840, followed by one at the University of Edinburgh in 1868.

The engineering chair at Glasgow was of particular importance for engineering education and for engineering in general, for from 1855 to 1872 it was occupied by W.J.M. Rankine. Rankine's influence extended well beyond the confines of Glasgow, for more than anyone else he was responsible for the creation and dissemination of a new branch of knowledge, engineering science.[11] Rankine's motivations were necessarily mixed. The entrenched science faculty had long exhibited a hostility to the intrusion of engineering professors into territory they claimed to be exclusively theirs. At the same time, Rankine was genuinely convinced that the progress of engineering depended on the development and application of a new form of science, one that was directly applicable to real-world engineering needs. Engineering science, as Rankine conceived it, would make optimal use of the theories and findings of pure science relevant to engineering problems, while at the same time integrating them with knowledge based on actual practice. It was a brilliant synthesis, used to excellent effect for such projects as improving the operating efficiency of steam engines and determining the actual stresses operating on a structure. Rankine's ideas were transmitted through textbooks such as *Manual of the Steam Engine and Other Prime Movers*, *Manual of Civil Engineering*, and *Manual of Appied Mechanics*. These books went through a vast number of editions and were used by engineering students for many years after the death of their author.

Rankine and his colleagues scored some impressive achievements, but Britain still lacked an adequate academic base for training engineers; at no time before the outbreak of World War I was there an enrollment of more than 1500 engineering students.[12] An apprenticeship continued to be the standard means of learning engineering, and even university graduates were not exempt. This form of education had its benefits, for it insured that students would get a lot of real-world experience. At the same time, however, it tended to lock British engineering into traditional, cut-and-try approaches to engineering that took little notice of emerging scientific knowledge. For some established British engineers this was all to the good; as Isambard Kingdom Brunel admonished one aspiring engineer,[13]

> I must strongly caution you against studying *practical* mechanics among French authors—take them for abstract science and study their statics, dynamics, geometry etc. etc. to your heart's content but never read their works on mechanics any more than you would search their modern authors for religious principles. A few hours spent in a blacksmith's and wheelwright's shop will teach you more practical mechanics. Read English books for practice. There is little enough to be learnt in them but you will not have to unlearn that little.

Some engineers undoubtedly benefited from a course of instruction that put practical experience ahead of book learning, but for many others an apprenticeship resulted in a fair amount of exploitation. An engineering apprenticeship could be an expensive proposition. An apprenticeship premium in civil engineering cost between 200 and 500 pounds during the nineteenth century,[14] making it prohibitively expensive for many young men of modest financial resources. And sad to say, some engineers were more interested in this payment than in educating their charges. At worst, an apprentice might be taken in by an engineer who was primarily interested in the premium, use him as cheap labor, and then "turn him loose at the end of his time with nothing more than a lukewarm testimonial."[15]

While a great amount of British engineering education was being conducted away from the universities, Germany was rising to technological pre-eminence through the establishment of first-rate engineering schools. Before 1870, a technical education geared to the needs of industry had suffered from low prestige, but during the last decades of the nineteenth century the *technische hochsculen* gained in rigor and status.[16] These institutes, most notably the Technical Institute of Munich (founded in 1868) and the Berline Technische Hochscule (founded in 1879), became models of advanced training, combining academic rigor with a responsiveness to the needs of a rapidly developing industrial economy. At the turn of the 20th century Germany had 25,000 science and engineering students in the universities to Britain's 3,000.[17] In 1910 there were 16,000 *technische hochschule* students, four times the number taking comparable courses in Britain.[18] The typically British way of training engineers had maintained the Industrial Revolution's momentum, but its deficiencies were all too evident by the beginning of the present century.

Engineering Education in the United States

As with France and Germany, a concern with national security and economic development provided the initial impetus for the formal education of engineers in the United States. The Revolutionary War had demonstrated that the new nation was seriously deficient in native engineering skills.[19] Washington's army had been heavily dependent on European military engineers, such as Louis Duportail (1743-1802), the chief engineer of the American army from 1777 to 1783, Thaddeus Kosciusco (1746-1783), and Louis de Tousard (1749-1817). Tousard went back to France after the war, but then returned to the United States where, it has been claimed, he served as the real founder of West Point.[20]

George Washington had vigorously pressed for the creation of a national military academy, whose principal function would be the education of engineers. This finally occurred in

1802, when the United States Military Academy at West Point was created by an act of Congress. It languished until 1812, when Sylvanus Thayer (1785-1872) became Superintendent and began the complete reorganization of the institution. Thayer was a graduate of Dartmouth College who had gone into the Army Corps of Engineers to teach mathematics at West Point. He brought in as Professor of Engineering Claudius Crozet (1790-1864), a graduate of the *Ecole Polytechnique* and an officer in Napoleon's army.[21] Crozet was instrumental in modeling West Point's engineering curriculum on that of the *Ecole Polytechnique*.

Much of the official support for the Military Academy came from government officials who expected that the skills acquired at the Academy would not be confined to the military realm, but would be useful in regard to "whatever respects public buildings, roads, bridges, canals, and all such works of a civil nature."[22] This was not a vain hope; between 1802 and 1837, 231 of the Academy's 940 graduates went into civilian engineering at some time in their careers.[23] We have already reviewed the careers of some of the most eminent of these engineers in Chapter 7.[24]

By the middle of the nineteenth century, West Point had lost its dominant position as college engineering departments and university-based schools of engineering were founded to meet the growing demand for formally trained engineers. Union College in Schenectady, New York established a course in civil engineering in 1845, while in nearby Troy, the Rensselaer School was reorganized in 1849 as the Rensselaer Polytechnic Institute, taking the *Ecole Centrale des Arts et Manufactures* as its model. At about the same time, engineering was finding its way into the curricula of Ivy League institutions, as Brown began to offer civil engineering courses in 1847 and Dartmouth established the Chandler Scientific School in 1851. Cornell set up the Sibley College of Engineering in 1868, and Yale's Sheffield Scientific School, founded in 1860, offered America's first course in mechanical engineering. Harvard graduated its first engineer in 1854, although its laggard performance (only 155 graduate engineers by 1892) was one of the motivations for setting up the Massachusetts Institute of Technology in 1861.

University-based engineering programs received their greatest boost in 1862 with the passage of The Morrill Land Grant College Act. It provided for the granting of 30,000 acres of federal land for each of a state's Senators and Representatives. The sale of these lands was then used to fund new institutions of higher education, which were required to offer instruction in "agriculture and the mechanic arts." Many of the newly created institutions were quick to establish schools of engineering; within ten years of the passage of the Morrill Act, the number of engineering schools had increased from 6 to 70.[25] Quantity was not always matched by quality, however; many of the new engineering schools were marred by incompetent faculty, ineffective teaching, and a lack of support by university administration.[26]

Private colleges of engineering and science also multiplied at this time. Cooper Union was founded in 1859 by the eccentric genius Peter Cooper, who thirty years earlier had built the steam locomotive "Tom Thumb" for trial on the Baltimore and Ohio Railroad. The Massachusetts Institute of Technology began instruction in 1861, although the Civil War prevented the graduation of its first class until 1869. Worcester Polytechnic Institute opened in 1865. The Stevens Institute of Technology, founded in 1870 by Robert Livingston Stevens, the third generation of an eminent family of engineers, became the first American institution to give a degree in mechanical engineering. To these pioneering institutions could be added the Case School of Applied Sciences (1880), Throop Polytechnic Institute (founded in 1891 and later renamed the California Institute of Technology), and the Carnegie Institute of Technology (1900).

As a result of this rapid development of engineering education, more and more practicing engineers had extensive academic preparation. The United States had 17 colleges of engineering in 1870, 85 in 1885,[27] and by 1890 the figure had risen to 110.[28] The number of students enrolled in engineering programs accelerated accordingly. In 1866 only about 300 engineers had graduated during the previous three decades.[29]

In 1890 there were 1000 students of engineering, and a decade later there were 10,000.[30]

This surge in the production of college-trained engineers marked a turning point for the profession. By 1880 academically trained engineers began to outnumber the bench-trained, although at this stage most of the college products would still have been in the junior ranks of engineering.[31] The change was gradual, and the impact of turning engineering into an occupation to be learned in the university was not fully felt until the twentieth century. Nevertheless, this nineteenth-century concern with education offers convincing evidence that the engineer was seen as a valuable and essential member of American society.

American engineering schools were often the site of a number of important educational innovations, such as the use of laboratory instruction as an essential part of the curriculum.[32] Of greater importance, American engineering schools and their curricula usually reflected the needs of the business firms that employed their graduates. Industry executives, many of them university-trained engineers, were often found on alumni boards and the education committees of professional societies, giving them a base from which they could exercise their influence over engineering education.[33] The result was a pattern of training that usually was quite responsive to the needs of prospective employers.

Even so, the connection between academia and enterprise could be fraught with tensions. This is exemplified by the experience of MIT's electrical engineering program, which under the leadership of Dugald Jackson (1865-1951) developed a cooperative program with General Electric to give students industrial experience alongside their regular course work.[34] The program, which ran from 1907 to 1932, was especially oriented toward preparing students to take on managerial roles, a logical ambition, given the usual career patterns of MIT graduates. But despite these good intentions and the basic soundness of the program, the cooperative idea encountered strong criticism from other engineering educators who decried the inattention to research and instruction in the fundamental underpinnings of engineering practice.[35] Nor was industry satisfied with the results; Jackson's program was criticized by his sponsors at General Electric who often voiced a concern that the engineers produced by the program were overqualified and were not well-prepared to meet very specific technical needs of the company.[36]

University-based engineering programs were also squeezed from another direction. Faculties in long-established disciplines often did not take kindly to the planting of engineering programs within the groves of academe.[37] Many engineering programs were kept out of the university mainstream, and often confined to separate faculties and even campuses. During the nineteenth century, engineering students, no matter what their academic qualifications and achievements, were excluded from membership in the academic honor society, Phi Beta Kappa. As a result, a separate engineering society, Tau Beta Pi, was founded in 1885.[38] Engineers might gain widespread respect for their contributions to the the nation's economy, but many within academia harbored a suspicion that the knowledge and practices that had made these contributions possible would pollute the rarified air of the scholarly community.

Educating Engineers in Asia

Just as Germany and the United States consciously developed engineering programs as instruments of technological and economic development, the governments of today's Third World nations have energetically sponsored the growth of engineering education. The rapid development of engineering education in conjunction with ambitious economic modernization programs can be seen in the world's most populous countries, India and China. In 1947, the year of India's independence from England, only 900 engineering students graduated from Indian universities; by the mid-1950s their number had tripled.[39] In the mid-1980s, India had more than 250,000 graduates of university engineering programs.[40] Similar strides were made in China after the establishment of the People's Republic in 1949. Before then, Chinese universities never enrolled more than 28,000 students of engineering, yet by 1958, 177,000 were enrolled.[41] So great was the emphasis on engineering that during the 1950s nearly a third of all university graduates were engineering majors.[42] Both countries also produced even larger numbers of graduates of technical schools, many of whom went on to work as engineers.

Impressive as these figures are, they obscure some deep-seated problems. India's production of engineers seems to have reflected the needs of educational institutions and not the economy's demand for university-trained engineers. As a result, nearly 19,000 graduate engineers found themselves unemployed in the mid-1980s.[43] According to its critics, India's top-heavy educational system has produced large numbers of poorly trained engineering graduates, while at the same time India's laggard industrial economy has failed fully to absorb even the competent ones.

China did not have the latter problem, for its centrally planned economy at least ensured that all engineering graduates would be placed in jobs. But the quality of many engineering graduates has been inferior to those who had been educated prior to 1949 or had been educated in the Soviet Union before the Sino-Soviet split in the early 1960s. Moreover, on several occasions the Chinese government has badly misused engineering personnel. Required participation in centrally directed political movements often detracted from the education and work of engineers. The most virulent of these, the Great Proletarian Cultural Revolution (1966-1976, according to the official chronology) created utter devastation. Engineers and other technical experts were often attacked. In the worst cases they were killed or driven to suicide, while others were often relegated to doing menial work. The universities were closed, often for several years, and when they reopened their curricula and recruitment policies were hardly congenial to high-quality engineering education.[44] Only after the death of Mao Zedong in 1976 and the accession to power of a pragmatic leadership bent on economic modernization did China once again become receptive to developing high quality engineering education. Yet this has created a new set of problems. China, like other Third-World nations, cannot always provide engineers with tasks commensurate with their education, skills, and ambitions, resulting in substantial numbers of underemployed engineers and a fair amount of political disaffection.

Without question, the non-Western nation with the greatest success in producing an able cadre of engineers has been Japan. When the Japanese government began aggressively to promote economic development after the Meiji Restoration (1868), it made the development of engineering skills one of the major components of its industrialization program. The Ministry of Public Works created the country's first college of engineering in 1873; four years later it became the Imperial College of Engineering. Under the leadership of 24 year-old Henry Dyer, a native of Glasgow, the college pioneered educational innovations such as project examinations and supervised work in attached factories and workshops. When Dyer and his associates returned home they took their ideas with them, and British engineering education began to profit from the experience that was gained in the course of creating a system of engineering education from scratch.[45]

Other colleges of engineering were formed within existing Japanese universities before the turn of the century, while at the same time a number of technical trade schools were created under state patronage.[46] By 1903 Japan had 240 technical schools of various degrees of quality in addition to university-based engineering and technical colleges.[47] These efforts continued through the century, and today Japan trails only the United States and the countries of the former Soviet Union in the number of university-trained engineers.

Professional Societies and the Upgrading of Engineering Education In the United States

Although engineering education developed rapidly during the second half of the nineteenth century, many practitioners continued to learn their skills in non-academic settings. In 1900, a year when 1000 engineering degrees were awarded, 43,000 engineers were already employed in the United States.[48] Even today, many working engineers have only minimal exposure to post-secondary education. According to the 1980 census, 35 percent of the self-reported engineers did not have a degree in engineering.[49] Of course, not all of the respondents claiming the title "engineer" do actual engineering work; moreover, some practicing engineers had degrees in related fields such as physics or chemistry. Even so, many engineers have in fact learned their skills on the job and have had little formal training in an engineering program.

The ability to assume the title "engineer" (and in fact do the work of an engineer) with only minimal educational qualifications continues to undermine the efforts of engineers who seek to assume the status of full-fledged professionals. As a result, much of the history of engineering education can be seen as an effort to raise engineering's status by requiring that practitioners acquire their knowledge and skills in a university setting.

In the United States the first systematic attempt to tightly couple engineering to a rigorous educational program came in 1893 with the formation of the Society for the Promotion of Engineering Education (later renamed the American Society for Engineering Education). In 1918 the Society, aided by the Carnegie Foundation for the Advancement of Teaching, issued a report that assessed the state of engineering education and gave some recommendations for the future. The report did not take issue with the prevailing idea that engineering was an essentially practical activity, or that engineering education should simulate actual work situations through laboratory work, industrial training, and shopwork.[50] At the same time, the report acknowledged the growing importance of the scientific foundations of engineering, and stressed that "...the college curriculum should aim to give a broad and sound training in engineering science, rather than a highly specialized training..."[51] Still, engineering practice could not rest on abstract knowledge selected by university faculty; the report also emphasized the importance of tying engineering education to the needs of employers.[52]

In the ensuing years the knowledge required of engineers was increasingly grounded in science and mathematics, thereby forcing some fundamental changes in the nature of engineering education. The need for rigorous training in mathematics and basic scientific principles was emphasized during the 1920s by another SPEE task force, this one headed by William Wickenden, an executive with The American Telephone and Telegraph Company. The Wickenden Report stressed that the "primary purpose of instruction in specialized engineering subjects should be to teach fundamental principles and methods rather than to train for particular kinds of work."[53] Many engineering schools heeded the call, and such staples of engineering education as shopwork, practical design work, and drafting began to be shouldered aside by an increased emphasis on physics, chemistry, and mathematics.

The rapid development of a number of engineering specialties — most notably electronics — strongly reinforced the belief that engineering required a sounder scientific foundation. Not surprisingly, the American Society of Electronic Engineers took the lead in stressing the scientific basis of engineering practice. One vociferous exponent of a scientifically based engineering education was Stanford's Dean of Engineering, Frederick Terman, who in a 1947 report to the president of his university warned that "Industrial activity that depends on imported brains and second-hand ideas cannot hope to be more than a parasite that pays tribute to its hosts, and is permanently condemned to an inferior competitive position."[54] Terman's experience guiding research projects during the World War II convinced him that American engineers had not risen to the challenge of the new electronic technologies, leaving the development of these fields to physicists and others with more rigorous training. Terman's concerns were also expressed in a 1951 committee report of the Engineering Council on Professional Development, which decried the watering-down of the engineering curriculum and the intrusion of courses in accounting and business; what would-be engineers needed was a solid foundation in basic science and engineering theory. As for the "art" of engineering, it was something best acquired "in the field."[55]

Other studies of engineering education (the Hammond Reports of 1940 and 1944, the Grinter Report of 1952-55, and the 1968 Walker Report) sounded the same theme. The Grinter report was of particular importance, for its recommendations became the basis for accrediting engineering schools. It called for even more emphasis on basic science and engineering science, and a consequent de-emphasis on practical engineering.[56] In Perrucci and Gerstl's summation, this report could "...be viewed as the culmination of efforts to move engineering from a practical art to a science-based profession..."[57]

The 1955 report of the ASEE's Committee for the Evaluation of Engineering Education went a step further by attempting to delineate a new discipline of "engineering science" that paralleled, but did not duplicate the basic sciences. The new discipline was an amalgam of a number of fields: the mechanics of solids (statics, dynamics, and strength of materials), fluid dynamics, thermodynamics, transfer rate mechanisms (heat, mass and momentum transfer), electrical theory (fields, circuits and electronics) and the nature and properties of materials.[58]

The two professional societies concerned with electronics, the American Society of Electrical Engineers and the Institute of Radio Engineers, continued to debate the optimal curriculum and pattern of instruction, and although Terman's views were never fully accepted, at least one branch of engineering indicated its determination to carve out a niche hitherto reserved for scientists.

Graduate Education in Engineering

Paralleling the "scientification" of the undergraduate curriculum was a concerted effort to add a postgraduate component to engineering education. Yale University was the first American institution to grant an engineering Ph.D in 1861, but the number of post-graduate degrees expanded very slowly at first. Many employers looked askance at the holders of master's degrees, believing that an advanced degree indicated only a student's unwillingness to face up to reality and take on responsibilities.[59] The increased technical demands of engineering in the post-World War II era helped to change this attitude. Before the war there were never more than 122 doctorates awarded in engineering in a given year, but since then more than 2000 have been awarded annually, with 3500 awarded in 1972 alone.[60] A much greater number received master's degrees; in 1980, 16,846 of these degrees were awarded—more than the total number of Japanese students *enrolled* in similar programs.[61]

Accompanying the growth of post-graduate education in engineering has been a growing commitment to coupling original research with advanced training. For the most part this has been a recent phenomenon. During the 1920s, for example, even the faculty of a branch of engineering as steeped in basic science as electrical engineering did little research. During 1920-25, the *Transactions* of the American Institute of Electrical Engineers contained an average of only nine university-originated articles per year. Seven of these nine articles were the product of five universities; the remaining institutions accounted for an average of less than two papers annually. An examination of the *Proceedings* of the Institute of Radio Engineers showed a similar pattern: an average of five publications per year of university origin, half of which came from physics departments. In fact, only one university engineering department could claim more than two publications during this period.[62]

The situation is dramatically different today. Engineering schools now support their programs and their faculty (as well as their legitimacy) by engaging in a great amount of research. More is involved here than the advance of knowledge; extensive research programs have changed the nature of engineering and engineering education. Research programs provide work support for graduate students who then join the ranks of the teaching faculty with little experience in actual engineering work. Thus the circle is closed, and the movement towards making engineering into a kind of science becomes self-perpetuating.[63] Where engineers were once instructed by teachers who had done actual engineering work, to an increasing degree they now are being taught by people with no such experience, but with a solid record of academic publications of sometimes questionable relevance to the actual work of engineering.

The Consequences of Modern Engineering Education

There can be little question that during the twentieth century a growing amount of technological change has been based on advances in scientific knowledge. Accordingly, the infusion of increasing amounts of science seems essential for an adequate engineering education. Yet doubts remain about the necessity and even the relevance of a curriculum heavily based on science.

Complaints about a "trained incapacity" of engineering school graduates have long paralleled the growth of science-based engineering curricula. As one turn-of-the century critic noted, "many men engaged in practical duties of an engineering nature frequently, and perhaps usually, complain that... young engineers almost invariably have failed to possess the ability to effectively address real-world engineering problems."[64] The increasingly intimate connection between science and technology has not completely dispelled these concerns. Even engineers themselves have on occasion expressed doubts about the relevance of their training. In his presidential address to the (British) Institution of Mechanical Engineers, Sir David Pye called into question the utility of theoretical training, laboratory experimentation, and the manipulation of formulas on paper. He further noted that "...by the time he has grown up as an engineer, indeed, his book learning will have become hardly more than a vaguely felt background to his experience, for the art of engineering includes a very great deal that lies outside of formulated theory. The rule of thumb plays, and will always play, an important part; and rules of thumb are accumulated by experience."[65]

Similar sentiments have been voiced by Howard Rosenbrock, who has taken note of the many aspects of engineering that are not amenable to scientific theories and procedures:[66]

> My own conclusion is that engineering is an art rather than a science and by saying this I imply a higher, not a lower status. Scientific knowledge and mathematical analysis enter into engineering in an indispensable way and their role will continually increase. But engineering also contains elements of experience and judgment and regards all social considerations and the most effective ways of using

human labour. These partly embody knowledge which has not been reduced to exact and mathematical form. They also embody value judgements which are not amenable to the scientific method.

Some business firms have been even harsher in their judgments, and on occasion have exhibited a marked hostility to college-trained engineers. An extreme example from many years ago is narrated by Christopher Hinton:[67]

> ...the university-trained engineer was unfashionable and was mistrusted in many branches of engineering industry throughout the 1920s and 1930s. Professor Inglis, under whom I had done research on the vibration of railway bridges during my last year at the university, told me that he had suggested to the Great Western Railway that they should take me back when I finished at Cambridge and had said that I was a good engineer. The railway company's answer was that "Hinton would have been a good engineer if he had stayed with us, but now that he has had three years at Cambridge, we wouldn't dream of taking him on again."

Such an attitude would be far less common today. Still, it is possible that there is more than a grain of truth in the belief that a university education, grounded in science, can distort a prospective engineer's training. The ultimate aim of engineering is to prepare students for the practical work of design, development, and testing. Yet as engineering education has become increasingly based on theory, mathematics, and science, students have begun to exhibit evident weaknesses. A 1986 National Research Council report noted that graduates of university engineering programs have become less capable of creating actual functioning systems and devices.[68] Training in basic science and mathematics has crowded out such areas as mechanical drawing and shop practices, formerly integral parts of the engineering curriculum. Above all, students are getting little hands-on experience with the actual craft of engineering. Most engineering graduates have learned how to analyze engineering problems but only rarely have they received any exposure to actual design work.[69]

With a great deal of contemporary engineering education centering on advanced mathematics, abstract concepts, and scientific theory, many of the firms employing graduate engineers find that their new employees require extensive on-the-job training before they can meet the needs of the company. And this feeling seems to be shared by many engineers themselves. In one British firm surveyed by Peter Whalley, three-quarters of the engineers felt that the applicable knowledge that they had learned in college could just as well have been learned on the job.[70]

It is also significant that in Japan, where a highly abstract and formalized system of instruction has long been the norm, on-the-job training is seen as an essential part of an engineer's education.[71] Many Japanese engineers end up in fields quite different from the ones in which they were trained. Surveys indicate that after two or three years on the job, more than 40 percent of Japanese engineers are pursuing technical specialties different from the one they had studied.[72]

The inability of significant numbers of engineering graduates to perform many traditional engineering tasks has stimulated the development of engineering technician programs, through which students are prepared for the practical tasks that still occupy the core of engineering practice.[73] These programs have grown rapidly. In the U.S. in 1983 nearly two hundred institutions offered 731 Associate or Baccalaureate programs in engineering technology, which enrolled more than 60,000 and 24,000 full-time and part-time students respectively.[74] Although these programs still trail regular engineering programs, which enrolled between 100,000 and 115,000 engineering students during the 1980s, in the U.S. they are taking on increasing significance in engineering education. In 1962 there was one 4-year engineering technology degree awarded for every 23 in engineering; by 1976 the figure was 1 to five, and rising.[75]

Engineering Education and the Quest for Professionalization

The expansion of scientific knowledge and its widespread application have exerted a profound influence on engineering education. But this is not the whole story. Changes in the style and content of engineering education also have been the result of the institutional needs of colleges and universities themselves. In turn, these institutional needs have reflected important changes that were taking place in the engineering profession itself.

As will be seen in greater detail in the next chapter, engineers have long been engaged in an effort to elevate engineering to full professional status. One of the hallmarks of a profession is that its practitioners possess and make use of a special body of knowledge. Not any sort of knowledge will do; it has to be structured and rather esoteric, the kind of knowledge that is best transmitted in a formal situation where the novice can learn from acknowledged masters.

To be sure, an apprenticeship program can be just such a setting, and for centuries aspiring engineers learned their trade by working alongside practicing engineers. In the United States, this was a situation that worked to the benefit of old-guard engineers, men who came from high-status families and were often owners or part owners of the firms in which they worked. Secure in their social and economic positions, they were not particularly concerned with elevating engineering to the status of a profession that could be put on the same footing as medicine or law.

In contrast, toward the end on the nineteenth century, the graduates and administrators of engineering schools tended to come from a different social background. Since they had no financial stake in the firms that employed them, their only hope of gaining the respect accorded to earlier generations of owner-engineers lay in their claim to being members of a recognized profession.[76] For this new breed of engineer, a thorough grounding in scientific knowledge was the key to professional status. The ability to learn and apply science and mathematics was not tied to prior social status. The engineering school thus could be a meritocracy that prepared talented and diligent students to enter the ranks of engineering.

Along with providing formal training in scientifically based principles of engineering, these schools attempted to cultivate in their students another aspect of professionalism: a broad sense of public responsibility.[77] One of the central tenets of professionals is that their actions are not selfishly motivated, but are inspired by a commitment to their clients and to society as a whole. Many engineering educators have concerned themselves with the development of ethical values. One example was Henry Eddy, who in his 1898 presidential address delivered before the Society for the Promotion of Engineering Education called for engineering schools to become more involved with the inculcation of professional ethics: "It remains then for engineering colleges to help organize the profession and to furnish the basis of such organization in a code of professional ethics which shall be worthy, unifying, and elevating."[78] This was more than a lofty sentiment; a professional code (and the public's belief in its operation) is essential if the members of an occupation hope to insulate themselves from external controls.

The infusion of large doses of science into the engineering curriculum has been driven by more than the objective requirements of good engineering. The possession of specialized knowledge is a key means of generating on-the-job autonomy and a way to bolster the engineer's claim to the status of the independent professional. With science enjoying great prestige in and out of academia, many engineers supported or at least acquiesced to the movement that sought to make engineering more like science. This effort may have enhanced the professional status of engineers; it is less certain that it has always resulted in higher quality engineering.

Notes

1 Frederick B. Artz, *The Development of Technical Education in France, 1500-1850* (Cambridge, Massachusetts and London: MIT Press, 1966), p. 99.

2 Ibid., p. 101.

3 Margaret Bradley, "Civil Engineering and Social Change: The Early History of the Paris *Ecole des Ponts et Chausees*," *History of Education*, 14 (1985), p. 176.

4 George S. Emmerson, *Engineering Education: A Social History* (New York: Crane, Russak, 1973), p. 77.

5 C.R. Day, "The Making of Mechanical Engineers in France: The Ecoles d'Arts et Metiers, 1803-1914," *French Historical Studies* 10, 3 (Spring 1978), p. 444.

6 Ibid., p. 452.

7 Artz, op. cit., pp. 247-53.

8 John Hubbel Weiss, *The Making of Technological Man: The Social Origins of French Engineering Education* (Cambridge, Massachusetts and London: MIT Press, 1982), p. 3.

9 Eric Ashby, *Technology and the Academics: An Essay on Universities and the Scientific Revolution* (New York: St. Martin's Press, 1959), p. 56.

10 Trevor I. Williams, *A Short History of Twentieth Century Technology* (Oxford: Clarendon Press, 1982), p. 3.

11 David F. Channell, "The Harmony of Theory and Practice: The Engineering Science of W.J.M. Rankine," *Technology and Culture* 23, 1 (January 1982)

12 R.A. Buchanan, "The Rise of Scientific Engineering in Britain," *British Journal for the History of Science* 18 (July 1985), p. 230.

13 Quoted in Ibid., p. 220.

14 W.J. Reader, *Professional Men: The Rise of the Professional Classes in Nineteenth Century England* (London: Weidenfeld and Nicolson, 1966), p. 122.

15 Ibid., p. 119.

16 Peter Lundgreen, "Education for the Science-Based Industrial State: The Case for Nineteenth Century Germany," *History of Education* 13, 1 (1984), p. 61. The *technische hochschule's* attainment of status equivalent to the university did not come easy, nor was its pre-eminent position in the training of German engineers always secure. See Kees Gispen, *New Profession, Old Order: Engineers and German Society, 1815-1914* (Cambridge: Cambridge University Press, 1989), pp. 78-80 and 150-59.

17 Gordon Roderick and Michael Stephens, "The Universities" in Gordon Roderick and Michael Stephens (eds.), *Where Did We Go Wrong? Industrial Performance, Education, and the Economy in Victorian Britain* (Lewes, Sussex: The Falmer Press, 1981), p. 200.

18 Paul L. Robertson, "Employers and Engineering Education in Britain and the United States, 1890-1914," *Business History* 23 (1981), p. 42.

19 L.P. Grayson, "The American Revolution and the 'Want of Engineers'," *Engineering Education* 75, 5 (February 1985)

20 Norman P. Wilkinson, "The Forgotten Founder of West Point," *Military Affairs*, 29, 4 (Winter 1960-61), pp. 177-88.

21 *A Biographical Dictionary of American Civil Engineers* (New York: American Society of Civil Engineers, 1972), pp. 30-31.

22 Daniel C. Calhoun, *The American Civil Engineer* (Cambridge, Massachusetts: MIT Press, 1960) p. 40.

23 Ibid., p. 43.

24 Although the pre-eminent role played by West Point in the education of engineers began to wane in the second half of the nineteenth century, the close association of engineering with military enterprises may be an enduring legacy. Some critics have noted that the military origin of engineering education is reflected in engineering's continued suffusion with the military virtues of order, hierarchy, and discipline. There has also been a longstanding complaint that involvement with military projects has resulted in a lack of concern for cost containment. Even in the 1830s one critic was asserting that "The habits, feeelings, and address of professed engineers is essentially military... [T]o excite the admiration of those whose tastes delight in elegance and grandeur without regard to cost is their chief ambition..." (Calhoun, op. cit., p. 137)

   If anything, this tendency has been reinforced in recent decades as military projects have taken a very large segment of engineering work. Defense-related engineering has done much to shape the curriculum, expand faculties, and provide a general orientation for the profession. Arthur L. Donovan, "Engineering in an Increasingly Complex Society," in National Research Council, Panel on Engineering Interaction with

Society, *Engineering in Society: Engineering Education and Practice in the United States* (Washington, D.C.: National Academy Press, 1985), p. 124.

25 David Noble, *America by Design* (New York: Knopf, 1977), p. 24.

26 Monte Calvert, *The Mechanical Engineer in America, 1830-1910: Professional Cultures in Conflict* (Baltimore: Johns Hopkins University Press, 1967), p. 48.

27 National Research Council, op. cit., p. 21.

28 Calvert, op. cit., p. 58.

29 L.P. Grayson, "A Brief History of Engineering Education in the United States," *Engineering Education*, 68 (December 1977), p. 250.

30 Edwin T. Layton, Jr., *The Revolt of the Engineers: Social Responsibility and the American Engineering Profession*, (Baltimore and London: Johns Hopkins University Press, 1986) p. 4.

31 John B. Rae, "Engineers Are People" *Technology and Culture*, 16, 3 (July 1975), pp. 415-17.

32 Grayson, "A Brief History," op. cit., p. 246.

33 Jeffrey K. Stine, "Professionalization vs. Special Interest: The Debate over Engineering Education in Nineteenth Century America," *Potomac Review* 27 (1984-85) p. 83.

34 A. Michal McMahon, *The Making of a Profession: A Century of Electrical Engineering in America* (New York: IEEE Press, 1984), pp. 70-79.

35 Ibid., pp. 76-77.

36 W. Bernard Carlson, "Academic Entrepreneurship and Engineering Education: Dugald C. Jackson and the MIT-GE Cooperative Engineering Course, 1907-1932)," *Technology and Culture*, 29, 3 (July, 1988)

37 Bruce Sinclair, "Inventing a Genteel Tradition: MIT Crosses the River," in Bruce Sinclair (ed.), *New Perspectives on Technology and American Culture* (Philadelphia: The American Philosophical Society, 1986)

38 National Research Council, op. cit., p. 21.

39 Humayun Kabir, *Education in the New India* (London: George Allen & Unwin, 1956), p. 16.

40 Dinesh Mohan, "Science and Technology Policy in India: Implications for Quality of Education," in Ratna Ghosh and Mathew Zachariah (eds.), *Education and the Process of Change* (New Delhi, Newbury Park, and London: Sage, 1987), p. 129.

41 Leo Orleans, *Professional Manpower and Education in Communist China* (Washington, D.C.: National Science Foundation, 1961), pp. 69-69.

42 Ibid., p. 127.

43 Mohan, op. cit., p. 129.

44 Jonathan Unger, *Education under Mao* (New York: Columbia University Press, 1982.

45 W.H. Brock, "The Japanese Connection: Engineering in Tokyo, London and Glasgow at the End of the 19th Century," *British Journal for the History of Science* 14, 48 (November 1981)

46 Mikio Sumiya and Keoji Taira, *An Outline of Japanese Economic History, 1603-1940: An Outline of Major Works and Research Findings* (Tokyo: University of Tokyo press, 1979)

47 William Wirt Lockwood, *The Economic Development of Japan: Growth and Structural Change* (Princeton: Princeton University Press, 1954), p. 511.

48 Robert Perrucci and Joel E. Gerstl, *Professionals without Community: Engineers in American Society* (New York: Random House, 1969), p. 59.

49 Robert Zussman, *Mechanics of the Middle Class: Work and Politics among American Engineers* (Berkeley and Los Angeles: University of California Press, 1985), p. 63.

50 Perrucci and Gerstl, op. cit., p. 59.

51 Ronald R. Kline, "Origins of the Issues," *IEEE Spectrum*, (November 1984), p. 39.

52 Noble, op. cit., pp. 203-6.

53 Kline, op. cit., p. 39.

54 McMahon, op. cit., p. 233.

55 Ibid., p. 235.

56 Kline, op. cit., p. 40.

57 Perrucci and Gerstl, op. cit., p. 60.

58 "Report of the Committee on Evaluation of Engineering Education," *Proceedings of the American Society for Engineering Education* 63, (1955-56) p. 37; quoted in Layton, op. cit., p. 95. See also Channel, op. cit., passim. This report did not sit well with Terman, one of the leading lights of electrical engineering, who was dismayed by the lack of stress on electronics, and feared that this vital area would be taken over by

"colleges of applied sciences", leaving engineers "to concentrate primarily on "dull trade school subjects." (McMahon, op. cit., p. 237)

59 Frederick E. Terman, "A Brief History of Engineering Education," *IEEE Proceedings*, 64, 9 (September 1977), p. 1402.

60 Zussman, op. cit., p. 9.

61 Earl H. Kinmonth, "Engineering Education and Its Rewards in the United States and Japan," *Comparative Education Review*, 30 (August 1986), p. 398.

62 Terman, op. cit., p. 1404.

63 Arnold D. Kerr and R. Byron Pipes, "Why We Need Hands-On Engineering Education," *Technology Review*, 90, 7 (October 1987), p. 38.

64 Noble, op. cit., p. 28.

65 Christopher Hinton, *Engineers and Engineering* (Oxford: Oxford University Press, 1970), p. 14.

66 Quoted in Mike Cooley, *Architect or Bee?* (Boston: South End Press, 1982), p. 106.

67 Hinton, op. cit., p. 4.

68 Kerr and Pipes, op. cit., p. 40.

69 Ibid. It is noteworthy that one survey found that engineers considered one of their most useful college courses to have been technical writing. They also noted that they would have profited from having had more courses centering on communication in general. (Gerstl and Hutton, op. cit., pp. 61-62) Another survey found that engineers selected the following subjects (from a list of 123) as the most likely to be helpful in one's career: (R. Richard Ritti, *The Engineer in the Industrial Corporation* (New York and London: Columbia University Press, 1971), p. 218)

Subject	*Percent indicating "would be helpful"*
management practices	65
technical writing	64
probability and statistics	60
public speaking	60
creative thinking	57
working with individuals	57
working with groups	55
speed reading	54
talking with people	53
business practices (marketing, finance, economics)	51

70 Peter Whalley, *The Social Production of Technical Work: The Case of the British Engineer* (Albany: State University of New York Press, 1986), pp. 55-56.

71 Po S. Chung, "Engineering Education Systems in Japanese Universities," *Comparative Education Review*, 30 (August 1986), p. 421.

72 Kinmonth, op. cit., p. 411.

73 Perrucci and Gerstl, op. cit., pp. 71-74. National Research Council, op. cit., pp. 13-14.

74 National Research Council, op. cit., pp. 31, 33.

75 Zussman, op. cit., p. 52.

76 Calvert, op. cit., p. 167.

77 Ibid., p. 276.

78 Stine, op. cit., p. 85.

Chapter 11

The Engineer Today: Origins, Roles, and Status

Today, more engineers are at work than ever before. Indeed, it is likely that more people are employed as engineers today than the total of all who worked as engineers prior to the mid-twentieth century. The great increase in the number of engineers along with ongoing developments in technology, organization, and government have markedly changed the nature of engineering. Today's engineer belongs to a lineage that includes Villard de Honnecourt, Han Gonglian, Leonardo da Vinci, John Smeaton, and all the other engineers we have met in the earlier pages of this book, but there is much that is different about engineering today. In this final chapter we will consider the origins of today's engineers, their continued struggles for recognition, some of the problems they face, and their search for professional status. The last topic occupies much of this chapter, for while it is important in its own right, it also helps us to tie a number of separate themes together.

On several occasions in this book the term "profession" has been used in describing the occupational status of engineers. But is this really an appropriate term? Are engineers entitled to the same occupational status as doctors and lawyers, or is it more appropriate to group them with social workers, nurses, school teachers and other occupations whose aspirations for professional status are yet to be realized? The issue is not simply one of terminology. Professions differ from other occupations in a number of significant ways; in particular, high occupational status along with a large measure of self-governance in work-related matters have set professionals apart from other workers. The effort to establish engineering as a profession encompasses most of the major issues surrounding the working life of today's engineers. But before taking up this matter, we need to consider the social origins of engineers, for the effort to enhance the occupational prestige of engineering reflects the fact that many engineers have entered engineering in the hope of enhancing their own position in society.

The Social Origins of Engineers

Occupation is the most important determinant of social position in modern societies. Through the attainment of a well-paying and prestigious occupation individuals can attain a status above the one held by their parents. This has been the experience of large numbers of engineers, as exemplified by Thomas Telford, Joseph Bramah, and the many other offspring of poor families who were able to convert their energies and talents into respected positions. To be sure, not all engineers have been upwardly mobile. A contrary example is Charles Parsons (1854-1931), who more than anyone else was responsible for making the steam turbine a practical source of power. He was the son of the third Earl of Rosse, a landed aristocrat and president of the Royal Society who produced the world's largest reflecting telescope and used it to discover spiral nebulae. The younger Parsons was educated at Cambridge, and served an apprenticeship at the Armstrong armaments works. His family's wealth was an important ingredient in his success as an engineer, for he spent over a thousand pounds of his own money while working on his turbine project.[1]

Few nineteenth century engineers came from such privileged circumstances, but many had solid middle and upper-middle class origins. The prospective engineers who were able to enter France's *Ecole Polytechnique* were rarely poor but talented boys who hoped to advance themselves through a career in engineering. Between 1830 and 1847, nearly 80 percent of *Polytechniciens* came from the families of high-ranking officials, military officers, professionals, and substantial merchants and manufacturers.[2] Other engineers enjoyed high status not so much because they were engineers, but because they owned the firms in which they practiced their craft. These individuals did not have to worry about their social status, for it was already secure. Instead, their concerns lay with preserving engineering as a "gentlemanly" occupation by ensuring that new engineers were drawn from social ranks similar to their own.[3]

These hopes were undermined by the steady growth of the number of engineers that occurred during the twentieth century, as well as by the widespread separation of engineering and entrepreneurial roles. As engineering expanded, relatively more engineers were drawn from the ranks of the middle and working classes. In the United States up to the late 1930s, nearly 70 percent of the graduates of engineering schools came from white-collar families, and 30 percent from blue-collar ones. The post-World War II era may have seen an increase in engineers drawn from more modest social circumstances; one survey conducted in the 1960s found that 60 percent of the graduates came from white-collar families and 40 percent from blue-collar ones.[4] Today, fewer engineers seem to be drawn from blue-collar families; one survey of the graduates of two university engineering programs in California found that nearly eighty percent of them had fathers who had been employed in professional, technical, managerial, or sales positions.[5] Still, many of these positions were undoubtedly relatively low in income and status, so for them and the remaining twenty percent of even more modest background becoming an engineer represented a distinct step upward. The upward mobility of significant numbers of engineers is not a distinctly American phenomenon. In the United Kingdom, 58 percent of university-trained engineers and nearly 80 percent of engineers without university degrees held a higher occupational rank than their fathers.[6]

Although the majority of engineering students are drawn from the ranks of the middle class, in most industrialized countries engineering continues to attract significant numbers of students from families headed by manual workers.[7] Engineering's attractiveness to young people from fairly modest circumstances is particularly evident when university engineering students are compared with students preparing for other careers. Students majoring in engineering come from families that on average are of lower social status than those of students majoring in medicine and law, business, biology, physical sciences, humanities, and the social sciences.[8] These educational patterns are replicated in the occupational positions achieved, for engineers have significantly more practitioners of working-class origins than medicine, law, and college teaching.[9]

Women Engineers

While engineering has provided an avenue of upward mobility for many young men, until recently it has not been of much significance for women in search of a well-paid, relatively prestigious occupation. An extreme example is West Germany, where only one of 500 engineers is a woman.[10] An engineering career has been more appealing to American women; in 1990, more than 11,000 women received bachelor's degrees in engineering—nearly 14 percent of total graduates.[11] This represents a very large increase in recent years, for in 1974 only 1.8 percent of engineering graduates were women. If this trend is maintained—and it must be noted that in recent years the enrollment of women in engineering programs has declined slightly—[12] the nearly complete hold on engineering by men will be diminished significantly.

It has been argued that the culture of engineering is a strongly masculine one that exhibits a marked hostility to women who dare to intrude into it.[13] During the early years of the twentieth century discrimination was overt; the American Institute of Electrical Engineers would accept women engineers only as a separate category without voting privileges or the right to hold office in the Institution.[14] Today, the barriers are more subtle. Women are able to do as well as men in university engineering courses; in fact, they often turn in superior performances. But once they are on the job, many encounter discrimination and overt harassment. Even when this is not the case, women engineers may find their careers impeded by a masculine engineering culture that emphasizes competitiveness, self-aggrandizement, and a strong interest in and aptitude for tinkering.[15] Large numbers of women engineers possess or have developed the intellectual skills and personality traits necessary to fit into this culture, but structural and attitudinal changes will be necessary if larger numbers of women are to go into engineering and develop successful careers within it.

Despite these obstacles, some women have persevered, and although relatively small in number, women engineers have made significant contributions in their fields. Their efforts have been especially important for the development of the computer. In the early nineteenth century Charles Babbage relied on Ada Lovelace to create the first programs for his mechanical computer. When computers became a practical reality in the 1940s, the first ENIAC program was written by Adele Goldstine, while Grace Hopper conceptualized and developed the first compiler program and applied it to the widely-used COBOL language.[16]

Edith Clarke (1883-1959) literally began her career as a "computer"—in the early twentieth century this meant a person who solved complex mathematical problems for scientific or engineering purposes. Clarke was first employed by American Telephone and Telegraph, where she did calculations essential for the operation of long-distance telephone lines. Although she already had a science degree from Vassar, she enrolled in engineering courses at the Massachusetts Institute of Technology, subsequently receiving the first electrical engineering degree awarded to a woman by that institution. After receiving her degree she had difficulty finding work as an engineer, so she again became a computer, this time for General Electric. While at GE she received a patent for a graphical calculator used for the mathematical analysis of electrical transmission lines; it generated solutions of line equations that encompassed inductance, capacitance, and distributed resistance. After a year's sojourn teaching physics in Turkey she returned to GE, this time as a full-fledged electrical engineer. She did important work in mathematical analysis of power systems, and became one of the first engineers to use a digital computer, in this case the pioneering differential analyzer created at the University of Pennsylvania. She also wrote a textbook on the analysis of AC circuits. With many accomplishments to her credit she left General Electric, and spent the last eleven years prior to her retirement teaching electrical engineering at the University of Texas.[17]

Perhaps the most influential American woman engineer was Lillian Moller Gilbreth (1878-1972). Although her contributions to industrial engineering are not as well recognized as those of her husband Frank, a good case can be made that in aggregate her achievements exceeded his. She outlived her husband by nearly fifty years, and did much to add an understanding of human behavior to the principles of Scientific Management. While her husband concentrated on rather narrowly conceived time-and-motion studies, Lillian Gilbreth greatly enlarged the scope of industrial engineering by introducing psychological concepts and empirical studies into its theory and practice. She also played a very significant role in establishing industrial engineering curricula into engineering programs throughout the United States and abroad. Finally, one of her books, *The Psychology of Management* was for decades the definitive treatise on the subject.[18] Although Lillian Gilbreth was not a typical engineer in that she was not engaged in the design of physical objects or the management of their construction and use, she was one of the most significant contributors to the field of industrial engineering. As such, she did much to enhance the role of engineers in the shaping of management theory and practice.

The Search for Professional Status

As we have seen, the participation of engineers in management has been of crucial importance in defining the careers of individual engineers and the status of engineering in general. Yet at the same time, the eagerness of many engineers to leave their field in order to join the ranks of management seems to indicate that engineering has not achieved an occupational status on par with the traditional professions. Although engineers generally enjoy relatively high incomes and occupational status, their position as true professionals remains problematic at best.

Engineering, it is true, shares some characteristics with established professions, most notably rigorous training and the application of theoretically grounded expertise. All professions distinguish themselves through their practitioners' knowledge and application of specialized information. Some engineering specialties owe their very existence to their ability to define a knowledge base unique to themselves. Chemical engineering emerged as a distinct segment of engineering through a conscious effort to generate and apply theories, concepts, and procedures that set it apart from chemistry and mechanical engineering. By invoking "unit operations" as they key conceptual element of chemical engineering, proponents of the emerging specialty were able to stake out their special turf and defend it against rival claimants.[19] This new sphere of knowledge was oriented least as much to professional needs as to technical ones. The essential components of unit operations—fluid flow, filtration, grinding, distillation, heat transfer, and so on—are primarily physical rather than chemical processes.[20] The important thing was their integration through the efforts of chemical engineers, thereby reserving for practitioners of this speciality a distinct niche in industry and academia.

Whatever its actual components, the distinctive knowledge acquired and used by engineers is at least potentially a solid source of professional status. A sizeable amount of engineering practice is based on science, which gives engineering an automatic claim to the validity of its knowledge base; from at least the nineteenth century, science has been seen as an objective and appropriate basis for authority in a democratic society.[21] Modern society accords great prestige to science, and as a result occupations that appear to be rooted in scientific knowledge have a strong claim to professional status.[22]

Still, the use of specialized, scientifically based knowledge does not in itself insure professional status for engineers. As a number of students of the professions have pointed out, the crucial element of any true profession is the ability of its members to govern themselves and exercise a significant amount of on-the-job discretion.[23] From this perspective, professional status rests on the ability of the membership to insulate themselves from outside sources of control. Members of the profession are deemed the only ones qualified to determine who is entitled to do the work, the procedures that result in their qualification, and the standards by which their work is judged.

According to these criteria, engineering still has not established itself as a profession of the traditional sort. As an editorial in a journal published by the Institute for Electronic and Electrical Engineers complained in 1969, the engineer's lack of "freedom of action" was the main reason that "engineering is not universally regarded as a profession."[24]

There is no escaping the fact that the engineer's desire for autonomy is compromised by the work environment of engineering. Engineering is rarely done as a solo activity. Rather, engineering has flourished as an occupation closely tied to the development of large organizations; as we have seen, the large corporation or government agency is the typical employer of engineers. Under these circumstances individual engineers have been able to pursue remunerative and often intellectually challenging careers, but they have not enjoyed an equal measure of on-the-job autonomy. Engineering faces an inevitable conflict between the professional's desire for independence and the requirement of loyalty to a bureaucratic organization.[25]

One crucial element of professional autonomy is the ability of a profession's members to police themselves. To cite two examples, doctors have been able to govern their own affairs by being the major participants in medical review boards, while college professors usually have prevented anyone else from determining and applying the standards that are used to evaluate their performance. Engineers have not enjoyed the same degree of success in governing themselves; as we have seen, engineers are subject to constraints created and applied by management. Individual engineers may play a large role in the oversight of engineering work, but they do so in a managerial capacity, not as engineers *per se*.

In all professions the exercise of autonomy is justified by the presumed competence, integrity, social responsibility of its practitioners.[26] And in fact, during the early decades of the twentieth centuries, the increasing bureaucratization of engineering led many young engineers to focus on public service as a way of enhancing their occupational status.[27] In recent years codes of engineering ethics have brought the issue of responsibility to the forefront by focusing on the engineer's need to work in accordance with the public interest.[28]

Translating a general desire to exercise social responsibility into a coherent set of values has been difficult. Engineering societies have encountered many difficulties in attempting to formulate a universal code of ethics that would encapsulate the social responsibilities exercised by engineers. In 1964 the Ethics of Committee of the Engineers' Council for Professional Development formulated three basic principles that it hoped would be adopted by all engineering societies: Engineers should be honest and impartial, advance the profession, and serve human welfare.[29] A more detailed prescription appeared in the code of ethics generated in 1977 by the Engineers Council for Professional Development:[30]

1. Engineers shall hold paramount the safety, health, and welfare of the public in the performance of their professional duties.

2. Engineers shall perform services only in the areas of their competence.

3. Engineers shall issue public statements only in an objective and truthful manner.

4. Engineers shall act in professional matters for each employer or client as faithful agents or trustees, and shall avoid conflicts of interest.

5. Engineers shall build their professional reputation on the merit of their services and shall not compete unfairly with others.

6. Engineers shall associate with only reputable persons or organizations.

7. Engineers shall continue their professional development throughout their careers and shall provide opportunities for the professional development of those engineers under their supervision.

It would be hard to take issue with these principles, but at the same time in their vagueness and generality they offer little guidance for engineers as they go about their everyday work. The most precise regulations governing an engineer's work continue to be stipulated by legal codes and contracts with employers.[31] Not only are ethical codes lacking in specificity, they do not have the power of sanctions undergirding them; few if any engineers have been censured by their professional society for harming the public health, welfare, and safety.[32] Yet there was one case in 1932 in which two engineers were expelled from the American Society of Civil Engineers for exposing the corrupt practices of another engineer. A court subsequently confirmed their accusations, but they still were not reinstated.[33]

With little practical guidance or ethical oversight from their professional societies, engineers have been unable to project the image of social responsibility that has allowed other occupational groups to claim the autonomy accorded to true professionals. On occasion, the attempt by an individual engineer to behave in a responsible manner has created a conflict with his or her employer, usually with unfortunate consequences for the engineer.

The Engineer as Whistle Blower

The problems inherent in attempting to exercise professional autonomy while serving as the employee of an organization emerge dramatically when an engineer draws attention to the negligence, mistakes, or malfeasance of his or her employer. In popular parlance, an engineer who goes to a government official, regulatory agency, or the media in order to point out errors or wrongdoings has "blown the whistle." Whistle blowers put their desire to serve the public ahead of loyalty to their employer, but in so doing they are likely to put themselves in a perilous position.

The most dramatic recent example of whistleblowing by an engineer followed the explosion of the space shuttle *Challenger* shortly after launch. A group of engineers from Morton-Thiokol, the manufacturer of the Shuttle's solid-fuel booster rocket, had warned the night before the launch that cold weather could cause a failure of the O-rings that sealed the segments of the booster rockets. After the erosion of an O-ring resulted in the tragedy that many had feared, Roger Boisjoly, a specialist in seals who had worked on NASA projects for twenty years, used his testimony before the official government board of inquiry to narrate the technical and managerial failures that led to the tragedy. This soon led to ostracism and isolation at Morton-Thiokol. Eventually diagnosed as suffering from traumatic stress, Boisjoly left his secure and remunerative job, sold his house, and left the community he had lived in for many years. Although he earned some income by giving lectures on the causes of the disaster, Boisjoly continued to suffer considerable financial and psychological disruption.[34]

Another well-publicized example of whistleblowing occurred in the San Francisco-Oakland bay area. Three engineers in the employ of the Bay Area Rapid Transit District (BART) repeatedly warned management about defects in the automatic train control system. Their concerns were not heeded, so they went over their superiors' heads and reported the problem to some members of BART's board of directors, who then informed the media without their consent. The result for the engineers was sadly predictable; they were asked to resign, and when they refused, they were summarily fired. They had to endure a prolonged period of unemployment, with attendant family and financial problems.[35]

Although he wasn't fired, W. Dan Deford, an electrical engineer for the Tennessee Valley Authority also paid a price for whistle blowing.[36] Troubled by TVA's possible omission of safety devices to measure the flow rate of diverted coolent in the event of a nuclear accident, Deford wanted to inform the

Nuclear Regulatory Commission of this possible deficiency. When his supervisor—also an engineer—refused, Deford told an NRC inspector of the potential problem during an on-site visit. For this he was publicly criticized and transferred to a non-supervisory position at another TVA facility. Deford was able to appeal his transfer to the Department of Labor because the authorization act for the Nuclear Regulatory Commission included an unusual employee protection provision that prohibited an employer from punishing an employee who has acted in support of certain health or safety regulations. The Department of Labor supported him by rejecting a subsequent appeal by TVA. The Department ruled that Deford's transfer was deliberate retaliation for whistle blowing, and recommended that he be returned to his original position, and awarded compensation for attorney's fees, stress-related medical bills, and $50,000 for damages to Deford's professional reputation (the Department of Labor subsequently voided the latter two awards.)

A less satisfactory outcome befell two employees who worked for a major defense contractor. After a chief engineer designed an aircraft disc brake that was woefully deficient in stopping power, he attempted to get his subordinates to rig tests and falsify data so that the brake's shortcomings would not be evident. When a lower-level engineer and a technical writer objected, they were told to keep their mouths shut and participate in the falsification. Subsequent testing by the Air Force demonstrated the brake's dangerous shortcomings. The engineer and technical writer informed the FBI of the falsification of the test results. Their employer designed a new brake, but avoided prosecution, successfully claiming that they had intended no deception in their reportage of the tests. Vilified by their supervisors, the two whistle blowers resigned under pressure. Meanwhile, all of the conspirators retained their jobs, except for two who were promoted to higher positions.[37]

In one of the instances narrated here, a whistleblower was vindicated and received at least partial compensation for his efforts to serve the public interest. In the other cases they lost their jobs or were demoted. Whatever the outcome, all of these cases show that a considerable amount of grief is likely to come to an engineer who attempts to invoke a professional code of ethics that puts the concerns of the public above those of the organization that employs him. It is therefore not surprising that many, if not most, engineers shy away from the prospect of putting their careers on the line in this manner. One survey of 800 randomly selected engineers conducted in 1972 found that nearly half of the respondents were inclined to "swallow the whistle"; that is, they felt that they were "restrained from criticizing their employer's activities or products." More ominously, more than ten percent believed that they were "required to do things which violated their sense of right and wrong."[38]

Another survey conducted in 1980 by *Chemical Engineering* indicated that working engineers are likely to have great difficulties in determining an ethical course of action even when they genuinely want to do the right thing. The survey presented a number of hypothetical cases involving such issues as falsifying data, rigging specifications, and adding minute amounts of potentially hazardous substances to food products. All of these cases were presented in such a way that a good case could be made for a number of different courses of action. In many of the cases, the responses spread over a considerable range, indicating that the respondents could not easily invoke clear ethical guidelines. Significantly, only a handful of the 43,000 respondents made any reference to the code of ethics of the American Institute of Chemical Engineers. Equally troubling, the respondents seemed to have little faith that their fellow engineers would behave in an ethical and responsible manner. Presented with one hypothetical case where a cyanide mixture was accidentally discharged into a sanitary sewer with no apparent ill effects, more than half of the respondents thought that they should report the incident to government authorities, as is required by law. But nearly three-quarters of the respondents stated that other engineers would either not report the incident or would leave it to management to determine if the authorities should be notified.[39]

Even in the absence of direct pressures, many engineers shrink from anything smacking of whistleblowing because their organizational environment creates subtle barriers to exercising responsibility. The organizational culture of engineering stresses the completion of projects and the deferment of responsibility for the project's consequences to management.[40] As one whistle-blowing engineer ruefully noted, "...executives set policy—more often than not by simply saying nothing—that communicates the idea to line employees that *broad questions of safety and usage are not their responsibility.*[41]

While the organizational context of engineering work seems to inhibit whistleblowing, some engineers can take some encouragement from government statutes that require the disclosure of dangerous activities or situations. Most notably, the Energy Reorganization Act of 1974 specifically mandates that directors or responsible officers of nuclear power plants must inform the Nuclear Regulatory Commission of failures to comply with safety standards or defects that could create substantial hazards.[42] The Civil Service Reform act of 1978 also includes a section that forbids reprisals by government officials taken against employees who disclose information "concerning the existence of any activity which the employee... reasonably believes constitutes... mismanagement... or a substantial and specific danger to the public health or safety."[43] In similar fashion, Congress has inserted "employee protection sections" in federal environmental and safety laws. These sections are designed to protect whistle-blowing employees from retaliation by requiring that the Labor Department provide hearings before outside examiners, as happened in the Deford case. Even so, the Labor Department has not demonstrated any great enthusiasm in protecting employees, failing even to notify employees of the protections available to them.[44]

Professional Societies and Professional Ethics

Caught between their employers and a generally indifferent government, whistle-blowing engineers are left with one source of organized support: their professional societies. Yet the protection of whistleblowers has been a task that most professional societies have been reluctant to take on. Many engineering societies still hold to the idea that their main purpose is the dissemination of technical information; the protection of their members is at best a secondary concern.

The reluctance of engineering societies to come to the defense of their members (and the public as well) is not simply the result of indifference or an excessively narrow definition of the role of a professional society. Since their inception engineering societies have been strongly influenced by members drawn from the ranks of management, who are naturally not predisposed to support the "disloyal" activities of their employees. Thus when the local chapter of the National Society of Professional Engineers came to the defense of the aforementioned BART engineers, some leaders of the state and central organization claimed that the local chapter's actions constituted an irresponsible attack on the engineer-managers in charge of the project.[45]

One of the most highly publicized cases of whistle-blowing centered on Ernest Fitzgerald, who was summarily fired after he called attention to the huge cost overruns on the Lockheed C5A transport plane. His professional association, the American Institute of Industrial Engineers, spurned Fitzgerald's request to "investigate the professional and ethical questions involved" on the grounds that it was a "technical organization" and not a "professional society." The influence wielded within the society by defense industry executives made this a predictable outcome.[46]

A somewhat better resolution was achieved by Charles Pettis. A civil engineer working on road construction in Peru, he pointed out to his employer, Brown and Root, that some building practices had potentially dangerous consequences. His employer attempted to silence him by offering him a choice of jobs anywhere in Latin America. When he refused, he was fired. His professional association, the American Society of Civil Engineers, did nothing for him, although its former executive director did conduct some inquiries and interceded with prospective employers on his behalf.[47]

Despite these shortcomings, engineering societies may be the engineer's only hope when he or she chooses to blow the whistle. Despite the negative examples noted above, some societies have assumed a more aggressive stance in these matters. The code of the Institute of Electrical and Electronic Engineers specifically states that "Members shall, in fulfilling their responsibilities to the community, protect the safety, health, and welfare of the public and speak out against abuses is those areas affecting the public interest."[48] In similar fashion, "The Engineer's Code" of the National Society for Professional Engineers states that:[49]

> The engineer will have proper regard for the safety, health and welfare of the public in the performance of his professional duties. If his engineering judgement is overruled by nontechnical authority, he will clearly point out the consequences. He will notify the proper authority of any observed conditions which endanger public safety and health... he will regard his duty to the public welfare as paramount.

In addition to lending moral support to whistle-blowing, some engineering societies have attempted to provide direct help for engineers who have engaged in whistleblowing. The Institute for Electrical and Electronic Engineers has been particularly noteworthy in this regard; it has provided legal aid and financial assistance to a few whistle-blowers, and has even instituted a special award for engineers who have demonstrated their ethical concerns through this action.[50] Even so, the way that the IEEE went about making its first award indicated a marked ambivalence within the society. The award ceremony, which honored the aforementioned BART engineers, was not on the official program of the Institute's annual meeting, and it received no publicity. On the other hand, a former president of the IEEE spoke at the ceremony while another sent a congratulatory message.[51]

These are important measures that deserve emulation by all professional societies. In addition to these efforts, the professional societies perhaps could do more to stimulate their members to pay more attention to their ethical responsibilities. The professional societies cannot create and oversee strict ethical guidelines that apply to each and every case. Rather than attempt to take upon themselves the task of trying to resolve the ethical problems encountered by their members, the major concern should be to stimulate thinking about ethical concerns and how they can be resolved at the individual level.

While taking individual responsibility for one's action is certainly important and laudable, there is still no getting around the fact that the great majority of engineers are employees, and as such are always subject to a great deal of organizational control. Under these circumstances, whistle-blowing will always be a hazardous venture that calls sharply into question the ability of the individual engineer to act according to the dictates of his or her own conscience. At the same time, the difficulties of the whistle-blowing engineer indicate how far engineers are from attaining the professional ideal of autonomous actors guided by an internalized sense of responsibility.

An Inclusive Occupation

The problems of the whistle blower demonstrate that while engineers may aspire to professional autonomy based on a transcendent code of ethics, there is no escaping the fact that they are subject to the constraints faced by all employees. Although efforts to secure professional status have become more problematic as employing organizations have grown in size and complexity, there is nothing fundamentally new about this situation. As one article bluntly stated in 1907, engineers had "forgotten that while lawyers work for clients, engineers work for employers.... The mechanical engineer and the electrical engineer are... essentially employees, and we believe that any attempt to lay down a comprehensive code of ethics to apply to these professions must be abortive."[52] Since that time, with more and more engineers working as employees of large organizations, the difficulty of exercising individual ethical judgment has become even more evident.

Another important component of professional independence entails the right to determine who is allowed to join the occupation and to set the standards determining their admission, as is the case with lawyers being admitted to the bar after passing an examination created and administered by the legal profession itself. There is no similar equivalent for engineers. Unlike medical and legal professional associations, engineering societies have not attempted to determine who is entitled to practice engineering through the establishment of mechanisms for qualifying engineers or otherwise restricting entry into the occupation.[53]

Some efforts to restrict entry into engineering ranks have occurred, but have met with limited success. In the United States, the National Society of Professional Engineers was established in 1934 to promote the certification of engineers and to stimulate the passage of laws to restrict entry into the occupation. Its efforts met with only partial success. Although some engineers have been duly certified by government-sponsored registration boards as being fit to practice engineering, registration is by no means universal. Only one-fourth of American engineers are currently registered, and this ratio is not likely to change in the future.[54] There is no need for most engineers to do so, for registration is required only for those engineers who work for local or state governments, or for consulting engineers, who sell their services directly to the public.[55] Most engineers apparently do not perceive registration/certification to be of any particular value; if professionalization is an important goal, it apparently will have to be pursued through other means.

British engineering societies have the greatest potential to control entrance into the profession, for they are unique in Europe in their right to determine professional qualifications.[56] British engineering societies have also worked with the government Board of Education to oversee engineering education and examinations. Even so, the use of educational standards to control entry into engineering has been only a secondary concern of the engineering societies.[57] In any event, British engineering societies can hardly be seen as successful proponents of professionalization, for the incomes and social status of British engineers have long been significantly lower than those of engineers in all other industrial nations.[58]

Engineering societies, far from promoting professional autonomy, have if anything contributed to the engineer's inability to assume true professional status. Part of the problem is that the societies cannot claim to represent the majority of engineers; in the U.S. no more than 25 percent of all engineers belong to national societies.[59] Moreover, the engineering societies have been at best lukewarm about pushing for registration and the consequent restriction of entry into engineering. In this, they reflect the nature of their membership. Many practicing engineers do not have university training; their engineering knowledge has for the most part been gained on the job. Under these circumstances, they are not likely to be enthusiastic about certification requirements that would likely stipulate a college diploma and/or passing an examination based on abstract knowledge.

At the other end of the membership spectrum, many engineering societies have members who are executives rather than practicing engineers. While some engineering societies have on occasion been rent by conflicts over the admittance of corporate executives,[60] the membership rolls of most societies include large numbers of executives who no longer work primarily as engineers, if they ever did. As might be expected, the corporations they represent have little interest in engineering society activities that might increase the autonomy of their employees.

With individual engineers apparently unconcerned about the attainment of professional status, and engineering societies doing little to change the situation, the strongest institutional push for professionalism has come from educational institutions.[61] Virtually from their inception, university-based engineering schools have attempted to increase the professional status of engineering, often in conflict with engineer-entrepreneurs whose occupational status was already guaranteed by their existing social status.[62]

It has often been remarked that the establishment of a university-based training program is an essential phase of an occupation's attempt to secure professional status. This may be so, but in the case of engineering at least, the impetus for doing so may come primarily from the university rather than from the practitioners themselves. While engineering school professors and administrators have been eager to bolster their own positions by emphasizing the scientific and theoretical basis of sound engineering practice, their efforts were not always welcomed by working engineers. Many engineers have been unwilling to recognize scientific training as the foundation of their craft.

The denigration of scientifically based engineering has a long history. In the dismissive words of one nineteenth century critic, "Scientific men from their manner of life, education and pursuits rarely possess the habits of economy, industry, foresight and judgment which are indispensable to great practical operations, and are more generally found among those whose experience and business knowledge are the result of a life of active industry and the basis of a sound judgment."[63] Another critic from the same era attacked scientifically inclined engineers for their "extravagant schemes [that are] accompanied with a long and abstruse treatise upon the theory of the matter, which no one but themselves can understand, in preference to the plain, practical and inexpensive plans of a less ostentatious individual."[64] Since these remarks were made, university-based educators and other proponents of scientifically based engineering have grown vastly in influence, although there still exists a considerable amount of tension between professors' and practitioners' conceptions of engineering.[65]

The Managerial Track as an Alternative to Professionalization

With engineering lacking some of the crucial characteristics of the established professions, many engineers seek wealth, status, and influence through entering the ranks of management. Engineers recognize that there is no escaping the organizational context of their work and that they must serve employers rather than individual clients. In so doing they tacitly acknowledge that their employers are the ultimate arbiters of their actions. This situation does not necessarily produce conflict. The goals of most engineers are congruent with those of their employers; they know that what counts is contributing to the success of their organizations, not the advancement of engineering for its own sake.[66] This orientation is demonstrated by one study of engineers working in the high-tech industries along Massachusetts' fabled Route 128; when asked if the pursuit of professional or organizational goals was more important for a successful career, only 9 percent opted for the former.[67]

Under these circumstances, attempts to create a "professional" niche for engineers unable or unwilling to enter management are not likely to be successful. Some firms have attempted to recognize the accomplishments of the technically skilled engineer who lacks an interest or aptitude for management through the establishment of a career track that parallels the managerial one. Unfortunately, where dual career ladders exist, they usually underscore the weak position of the engineer who attempts to define success in terms of technical achievements alone. In Goldner and Ritti's words, dual career ladders are largely management devices that serve to "maintain commitment on the part of those specialists who would ordinarily be considered failures for not having moved into management. Identification as a professional has become a way to redefine failure as success."[68]

When movement up the managerial ladder is the ultimate standard of occupational achievement, there is little hope that engineers will be able to define themselves according to the criteria of the professional. In Layton's summation, "In the long run, the most effective check on professionalization by business has been a career line that carries most engineers into management."[69] The attainment of professional status is always a difficult effort that requires a great deal of collective action; for most engineers attainment of a managerial position is a far more attainable goal, and one that can be pursued in an individualistic fashion.

Facing the Future

With the exception of the women engineers whose accomplishments have been briefly recounted, the engineers featured in this chapter have either been victimized whistle blowers or largely anonymous participants in inconclusive struggles over professional status. In similar fashion, the chapters on organization and management have given some indication of the extent to which engineering has become embedded in large, complex organizational structures. After reading these chapters one might infer that engineering has ceased to be a heroic occupation, one far removed from earlier, more glorious epochs when engineers built cathedrals, created railroad systems, and designed machine tools that worked with amazing precision. This is not our intent. Unquestionably, the activities, roles, and statuses of engineers have been influenced by changes in technology, organization, and the political and economic environment. But there is still a place for individual brilliance. It is well to remember that the term "engineer" springs from the same linguistic source that has given us the word "genius." Today's successful engineer needs the same blend of knowledge, perseverance, and creativity that in times past generated engineering accomplishments great and small. The need for the engineer's talents has not decreased; engineers will have to make use of all of their genius if humanity is to meet the challenges of pollution, overpopulation, diminishing natural resources, and the provision of a decent standard of living for the world's poor. It is to be hoped that this book has helped to illuminate the lives and achievements of some of history's great engineers; without their efforts the world would be a poorer place materially and culturally. At the same time, we hope that we have shown that while engineers have played a powerful role in the shaping of their societies, they are themselves the products of particular times and places.

Notes

1. W. Barrett Scaife, "The Parsons Steam Turbine," *Scientific American* 252, 4 (April 1985)

2. John H. Weiss, "Bridges and Barriers: Narrowing Access and Changing Structure in the French Engineering Profession, 1800-1850," in Gerald L. Geison (ed.) *Professions and the French State, 1700-1900* (Philadelphia: University of Pennsylvania Press, 1984), p. 43.

3. Monte Calvert, *The Mechanical Engineer in America, 1830-1910: Professional Cultures in Conflict* (Baltimore: Johns Hopkins University Press, 1967), pp. 8 and 131.

4. Carolyn Cummings Perrucci, "Engineering and the Class Structure," in Robert T. Perrucci and Joel E. Gerstl (eds.), *The Engineers and the Social System* (New York: John Wiley and Sons, 1969), pp. 283-84.

5. Judith S. McIlwee and J. Gregg Robinson, *Women in Engineering: Gender, Power, and Workplace Culture* (Albany: State University of New York Press, 1992), p. 28.

6. Joel E. Gerstl and Stanley P. Hutton, *The Anatomy of a Profession: A Study of Mechanical Engineers in Britain* (London: Tavistock, 1966), p. 24.

7. Wouter Van den Berghe, *Engineering Manpower: A Comparative Study of the Employment of Graduate Engineers in the Western World* (Paris: UNESCO, 1986), p. 45.

8. Robert L. Eichorn, "The Student Engineer," in Robert T. Perrucci and Joel E. Gerstl, *The Engineers and the Social System* (New York: John Wiley and Sons, 1969), p. 124.

9. Carolyn Perrucci, op. cit., pp. 282-237.

10. Stanley Hutton and Peter Lawrence, *German Engineers: The Anatomy of a Profession* (Oxford: The Clarendon Press, 1981), p. 35.

11. *Scientific-Engineering-Technical Manpower Comments*, 29, 3 (April-May 1992), p. 26.

12. National Research Council, Panel on Technology Education, *Engineering Technology Education* (Washington, D.C.: National Academy Press, 1985), p. 55.

13 For an analysis of the male-centeredness of traditional engineering, see Sally Hacker, "The Culture of Engineering: Women, Workplace, and Machine," in Joan Rothschild (ed.), *Women, Technology, and Innovation* (New York: Pergamon, 1982). For a discussion of how early twentieth century engineering was identified with masculinity, martial prowess, and imperialism, see Elizabeth Ammons, "The Engineer as Cultural Hero and Willa Cather's First Novel, *Alexander's Bridge*," *American Quarterly*, 38 (Winter 1986).

14 A. Michal McMahon, *The Making of a Profession: A Century of Electrical Engineering in America* (New York: IEEE Press, 1984), p. 58.

15 McIlwee and Robinson, op. cit.

16 John D. Ryder and Donald G. Fink, *Engineers and Electrons: A Century of Electrical Progress* (New York: IEEE Press, 1984), p. 179.

17 James E. Brittain, "From Computer to Electrical Engineer: The Remarkable Career of Edith Clarke," *IEEE Transactions on Education* E24, 4 (November 1985)

18 Martha M. Trescott, "Women Engineers in History: Profiles in Holism and Persistence," in Violet B. Haas and Carolyn Perrucci (eds.), *Women in Scientific and Engineering Professions* (Ann Arbor: University of Michigan Press, 1984), pp. 192-202.

19 Terry Reynolds, "Defining Professional Boundaries: Chemical Engineering in the Early 20th Century," *Technology and Culture*, 27, 4 (October 1986), pp. 707-12.

20 William F. Furter, "Chemical Engineering and the Public Image," in William F. Furter (ed.), *A Century of Chemical Engineering* (New York: Plenum, 1982), p. 397.

21 Burton Bledstein, *The Culture of Professionalism: The Middle Class and the Development of Higher Education in America* (New York: W.W. Norton, 1976), pp. 88-92.

22 Harold L. Wilensky, "The Professionalization of Everyone?" *American Journal of Sociology* 70, 2 (September 1964), p. 138.

23 Terence J. Johnson, *Professions and Power* (London: Macmillan, 1972)

24 McMahon, op. cit., p. 258.

25 Edwin T. Layton, Jr., *The Revolt of the Engineers: Social Responsibility and the American Engineering Profession*, (Baltimore and London: Johns Hopkins University Press, 1986), p. 1.

26 Robert Dingwall and Philip Lewis, *The Sociology of the Professions* (London: Macmillan, 1983), p. 41.

The Engineer Today 259

27 Bruce Sinclair, *A Centennial History of the American Society of Mechanical Engineers, 1880-1980* (Toronto: University of Toronto Press, 1980), p. 99.

28 Stephen H. Unger, *Controlling Technology: Ethics and the Responsible Engineer* (New York: Holt, Rinehart, and Winston,1982), p. 139.

29 William G. Rothstein, "Engineers and the Functionalist Model of Professions," in Perrucci and Gerstl, op. cit., p. 89.

30 Roger Mayne and Stephen Margolis, *Introduction to Engineering* (New York: McGraw-Hill, 1982), p. 44.

31 Kenneth Prandy, *Professional Employees: A Study of Scientists and Engineers* (London: Faber and Faber, 1965), p. 71

32 Unger, op. cit., p. 56.

33 Ibid., p. 47.

34 The events leading to the *Challenger* disaster and its consequences are narrated in Trudy E. Bell and Karl Esch, "The Fatal Flaw in Flight 51-L," *IEEE Spectrum* (February 1985).

35 Robert Perrucci, Robert M. Anderson, Dan E. Schendel, and Leon Trachtman, "Whistle-Blowing: Professionals' Resistance to Organizational Authority," *Social Problems*, 28, 2 (December 1980)

36 Rosemary Chalk, "The Miners' Canary," *The Bulletin of the Atomic Scientists* 38, 2 (February 1982), pp. 17-18.

37 Frank Vandiver, "Why Should My Conscience Bother Me?", in Robert Heilbroner, et al., *In the Name of Profit* (New York: Doubleday, 1972), pp. 3-31.

38 James Olson, "Engineer Attitudes Toward Professionalization, Employment, and Social Responsibility," *Professional Engineer*, August 1972.

39 Philip M. Kohn and Roy V. Hughson, "Perplexing Problems in Engineering Ethics," *Chemical Engineering*, 87, 9 (May 5, 1980) and "Ethics Survey Report," *Chemical Engineering* 87, 19 (September 22, 1980)

40 Chalk, op. cit., p. 20.

41 Peter Faulkner, "Exposing Risks of a Nuclear Disaster," in Allen F. Westin, (ed.), *Whistle-Blowing: Loyalty and Dissent in the Corporation* (New York: McGraw-Hill, 1981), p. 52. Emphasis in original.

42 Peter Raven-Hansen, "Dos and Don'ts for Whistleblowers: Planning for Trouble" *Technology Review*, 82, 6 (May 1980), p. 36.

43 Rosemary Chalk and Frank von Hippel, "Due Process for Dissenting 'Whistle-Blowers'," *Technology Review*, 81, 7 (June/July 1979), p. 54.

44 Ibid.

45 Ibid., pp. 52-53.

46 Ralph Nader, Peter J. Petkas, and Kate Blackwell, *Whistle- Blowing: The Report on the Conference on Professional Responsibility* (New York: Grossman, 1972), p. 52.

47 Ibid., pp. 135-39.

48 Chalk, op. cit., p. 19.

49 Chalk and von Hippel, op. cit., p. 50.

50 Chalk, op. cit., p. 21.

51 Unger, op. cit., p. 67.

52 Calvert, op. cit., p. 267.

53 Prandy, op. cit., p. 72.

54 Rothstein, op. cit., p. 85.

55 Ibid., p. 83.

56 Hutton and Lawrence, op. cit., p. 71.

57 J.O. Marsh, "The Engineering Institutions and the Public Recognition of British Engineers," *International Journal of Mechanical Engineering Education* 16, 2 (1987), p. 123.

58 Van den Berghe, op. cit., p. 197.

59 Bruce Sinclair, "Local History and National Culture: Notes on Engineering Professionalism in America," *Technology and Culture*, 27, 4 (October 1986), p. 684.

60 McMahon, op. cit., pp. 117-24.

61 Calvert, op. cit., p. 167.

62 Calhoun, op. cit., p. 137.

63 Ibid., p. 188.

64 Ibid.

65 Eliot Freidson, *Professional Powers: A Study of the Institutionalization of Formal Knowledge* (Chicago and London: University of Chicago Press, 1986), pp. 82-83.

66 R. Richard Ritti, *The Engineer in the Industrial System* (New York and London: Columbia University Press, 1971), pp. 48-50.

67 Paula Leventman, *Professionals Out of Work* (New York: Basic Books, 1981), p. 86.

68 Fred H. Goldner and R.R. Ritti, "Professionalization as Career Immobility," *American Journal of Sociology*, 72 (March 1976), p. 490.

69 Layton, op. cit., p. 8

Index

Name Index

Agrippa, Marcus Vipsanius, 20-21, 27
Agricola, Giorgius (Georg Bauer), 73-74
Alberti, Leon Battista, 61-62
Alexander the Great, 12
Ammann, Othmar, 161
Andreossy, Francois, 78
Anthemios of Tralles, 39
Apollodurus of Damascus, 24
Appius Claudius, 19
Archemedes, 13-14, 27, 52
Aristotle, 10
Armstrong, Edwin, 168-69

Babbage, Charles, 239
Baldwin, Laommi Sr., 139-140, 141
Baldwin, Laommi Jr., 140, 148
Baldwin, Matthias, 152
Banaker, Benjamin, 141
Bentham, Sir Samuel, 123
Bi Lin, 19
Biringuccio, Bannoccio, 74
Black, Joseph, 95, 116, 119
Boisjoly, Roger, 245
Boulton, Matthew, 118, 119-21
Boydon, Uriah, 152
Bradeley, Humphrey 74-75
Bramah, Joseph, 122-23, 236
Bridgewater, Duke of, 91
Brindley, James, 91, 97, 139
Brunel, Isambard Kingdom 103-105, 140, 211
Brunel, Marc Isambard, 103, 123-125, 139
Brunelleschi, Filippo, 59-60

Calley, Joseph, 114-115
Carnot, Lazare, 86-87, 89, 98

Cavendish, Henry, 120
Chao Kuo, 18
Clarke, Edith, 239
Cleon, 15-16
Clinton, DeWitt, 3, 20, 144
Colbert, Jean-Baptiste, 78
Cooper, Peter, 214
Corliss, George H., 151
Cosnier, Hugues, 77
Coulomb, Charles Agustin, 86, 89
Crozet, Claudius, 213
Ctesibius (Ktesibius), 12-13, 16, 26
Cugnot, Nicholas-Joseph 99-100

Darwin, Erasmus, 119
daVinci, Leonardo, 26, 58, 62-65, 66, 123
Davy, Sir Humphrey, 101
deBelidor, Bernard Forest, 88, 98
deBethune, Maximilien, duc de Sully, 75-76
deCraponne, Adam 76
Deford, W. Dan, 245, 248
deHonnecourt, Villard, 41, 45-46
deSaint-Simon, Henri Comte, 196
deTousard, Louis, 212
deVauban Sebastian le Prestre, 86, 89, 98
Dick, Robert, 116
di'Dondi, Giovanni, 49
diGiorgio, Francesco, 65-66
diNovate, Bertola, 66-67
Du Shi, 18
Duportail, Louis, 212
Dyer, Henry, 218

Eads, James B., 150-51
Eddy, Henry, 227
Edison, Thomas, 138, 164
Eilmer of Malmsbury, 46-47

Ellet, Charles Jr., 146-47
Ellicot, Andrew, 141
Eupalinos, 11-12
Evans, Oliver, 100, 141-42

Fairbairn, William, 128, 129
Fioravanti, Ridolfo, 66
Fitzgerald, Ernest, 249
Fontana, Domenico 67-68
Francis, James Bicheno, 151-52

Gantt, Henry L., 195
Geddes, James 4, 145, 146
Geng Xun 19
Gilbreth, Lillian Moller, 240
Goldstine, Adele, 239
Grundy, John, 93

Hadrian, 24
Han Gonglian, 49
Han Qi, 19
Harapolos, 12
Haupt, Herman, 150, 186
Hennebique, Francois, 161
Herodotus, 11-12
Heron (Hero), 16, 45, 60
Hieron of Syracuse, 13, 26
Holley, William, 186
Hopper, Grace, 239
Horikoshi, Jiro, 171-72
Hornbolower, Jonathan, 120
Hornblower, Jonathan Carter, 120

Iktinos, 11
Imhotep, 8-9
Ineni, 9-10
Isidoros of Miletus, 39

Jackson, Dugald, 215-16
Jervis, John B., 145-46, 152, 186
Johnson, Clarence ("Kelly"), 169-70
Johnson, Samuel, 95
Julius Caesar, 20

Kallikrates, 11
Kettering, Charles, 165, 174-75
Knight, Jonathan, 143

Kosciusco, Thaddeus, 212
Kyeser, Konrad, 51

Latrobe, Benjamin Henry, Sr., 147-48, 186
Latrobe, Benjamin Henry, Jr., 148
Lavoissier, Antoine, 89
Lenin, V.I., 192-93
Li Bing, 17-18
Lincoln, Abraham, 150
Lintlaer, 75
Locke, Joseph, 105
Long, Stephen H., 143
Lovelace, Ada, 239

McAdam, 87, 95, 98
McCallum, David C., 186
McClellan, George B., 186
McNeill, William Gibbs, 143-44, 151

Maillart, Robert, 161-62
Mao Zedong, 217
Marvel, Andrew, 73
Maudsley, Henry, 122-23, 125-26, 128, 129
Maxim, Hiram, 167
Meda, Giuseppe, 67
Metcalf, John, 91-92, 97
Monge, Gaspard, 86, 89, 98, 208
Murdock, William, 100, 120
Myddleton, Sir Hugh, 72-73, 89

Nasmyth, James, 125-127, 129
Nekhebu, 9
Newcomen, Thomas, 114-15
Nonius Datus, 20

Paine, Thomas, 98
Papin, Denis, 113
Parsons, Charles, 236
Perronet, Jean-Rodolphe, 87
Peter of Colechurch, 39
Pettis, Charles, 249
Philon of Byzantium, 13, 16
Pitot, Henri, 88, 98
Plato, 10, 196
Plutarch, 13-14

Index

Pollo, Marcus Vitruvius, 22-24, 29, 45, 61
Polo, Marco, 47
Priestley, Joseph, 119
Ptolemy Philadephos, 14-15

Ramelli, Agostino, 69
Rankine, W.J.M., 210-11
Rennie, George, 99
Rennie, John Sr., 39, 94, 95-96, 97, 98, 99
Rennie, John Jr., 72, 96
Riquet de Bonrepos, Pierre-Paul, 78-79
Roberts, Richard, 125-26, 129
Roebling, John Agustus, 160-61
Roebling, Washington, 160-61
Roebuck, John, 118
Robinson, Moncure, 147, 186
Robison, John, 95, 96, 116-17

San Gallo, Antonio, 67
San Gallo, Francesco Giamberti, 67
San Gallo, Giuliano, 67
Savery, Thomas, 113-114, 129
Scott, Howard, 197
Sellers, William, 153-54
Senamut, 9
Sennacherib, 7-8
Sextus, Julius Frontinus, 21-22, 28
Sforza, Francesco, 58
Sforza, Ludovico, 62
Shakespeare, William, 2
Shi Lu, 18
Smeaton, John, 92, 96, 115, 121-22, 148
Smiles, Samuel, 111
Smith, Adam, 117
Socrates 10
Sostratos of Knidos, 28
St. Benezet, 39
Stalin, Josef, 190, 191
Stanley, William, 167-68
Steinmetz, Charles, 197
Stephenson, George, 101-103
Stephenson, Robert, 101-103, 105, 144
Stevens, Robert Livingston, 214

Stevenson, David, 149-50, 156 n.24
Stevin, Simon, 70-71
Stone, Charles A. 184
Su Song, 48-49

Taccola (Mariano di Jacopo), 60-61, 69
Tartaglia, Nicolo, 68-69, 70, 88
Taylor, Frederick W., 154, 192-95
Telford, Thomas, 87, 94-95, 97, 98-99, 144, 236
Terman, Frederick, 220
Tertullian, 1-2
Thayer, Sylvanus, 213
Thomson, J. Edgar, 186
Thomson, William, Lord Kelvin, 106 n.7
Tresauguet, Pierre-Marie-Jerome, 87, 95
Trevithick, Richard, 100-101

Vaspasian, 25
Veblen, Thorstein, 196-97
Vermuyden, Cornelius, 71-72, 89
Vitruvius, see Pollo, Marcus Vitruvius

Washington, George, 212
Watt, James, 96, 100, 115-121, 129
Webster, Edwin S. 184
Wedgwood, Josiah, 119
Wernwag, Lewis, 139
Westinghouse, George, 167
Weston, William, 139, 140
Whistler, George Washington, 143-44
White, Canvass, 4, 145, 146
Whitney, Willis, 164
Whitworth, Joseph, 125, 127-28, 153
Wickenden, William, 220
Wilkinson, John, 118, 125
William of Sens, 40
Woolf, Arthur, 120
Wright, Benjamin, 4, 145, 146

Yeoman, Thomas, 93
Yi Xing, 48

Xerxes, 12

Subject Index

Acqueduct of Nimes, 24, 28
Aeolipile, 16, 113
Aesthetics, 11, 59, 129, 161
American Institute of Chemical Engineers, 187, 247
American Institute of Electrical Engineers, 221, 222, 238
American Institute of Industrial Engineers, 249
American Society of Civil Engineers, 151, 244, 249
American Society of Electronic Engineers, 220
American Society for Engineering Education, 219
American Society of Mechanical Engineers, 150, 187, 193
Apprenticeship, 59, 70, 72, 90, 91, 92, 98, 102, 103, 105, 116, 120, 125, 126, 128, 138, 139, 211-212, 226
Autonomy, 104, 127, 140, 227, 241-243

Bell Laboratories, 165
Briare Canal, 77
Bridges, 1, 12, 20, 32 n.48, 39, 58, 60, 62, 75, 95, 102-103, 147, 150, 160-62, 213
British Institution of Civil Engineers, 97, 103, 144
British Institution of Mechanical Engineers, 102
Bureaucratization, 172-75

Caledonian Canal, 93
Canal du Midi, see Languedoc Canal
Canals (see also specific canals), 7, 10, 15, 17, 18, 63, 66-67, 75, 76, 90, 91, 138, 139, 140, 144, 152, 213
Cathedrals, 37, 39, 40-45, 59-60
Certification, 252-53
Charolais Canal, 76
Codes of Ethics, 243-344, 247-251

Corps des Ingenieurs du Genie Militaire, 86, 87, 97
Corps des Ponts et Chausees, 87, 88, 97
Corvee Labor, 87
Cost Accounting, 187-88

De-skilling, 175-77
Dual Career Ladders, 176-77, 255
Dupont, 186

Ecole Centrale des Arts et Manufactures, 209, 213
Ecole des Ponts et Chausees, 147, 207, 208
Ecole du Corps Royale du Genie, 207-208
Ecole Polytechnique, 89, 208, 213, 236
Ecoles d'arts et metiers, 208-209
Eddystone Lighthouse, 92-93
Education, 3, 43, 58, 66, 70, 86, 90-91, 92, 94, 95, 102, 105, 112, 114, 116, 125, 126, 128, 130, 139-40, 142, 143, 145, 146, 147, 148-49, 150, 151, 171, 207-233, 253-54
Engineering Associations, 93, 97, 219-222, 227, 243, 244, 248-53
Erie Canal, 3-4, 20, 66, 78, 139, 144-45, 147, 159-60

Finances, 23, 28, 60, 68, 90, 105, 118, 168, 174-75, 187
Franklin Institute, 153-54

General Electric, 163, 164, 168, 184, 215-216, 239-240
General Motors, 163, 165, 174
Great Eastern, 104
Grinter Report, 220-221

Institute for Electronic and Electrical Engineers, 242, 250
Institute of Radio Engineers, 221, 222

Kanawha Canal, 147

Languedoc Canal (Canal du Midi), 77-79, 88, 91
Lighthouse of Alexandria, 28
Literacy, 44, 52, 79
Lockheed "Skunk Works", 169-70
Lunar Society of Birmingham, 119-120

Machine Tools, 111, 121-131, 153-54
Mathematics, 10, 13, 23, 39, 44-45, 52, 58, 62, 68-69, 70-71, 86, 112, 146, 208, 209, 219-220, 224, 226
Measuring Instruments, 27
Middlesex Canal, 140
Mining, 25, 73-74, 100-101, 113
Morrill Land Grant College Act, 214

National Society of Professional Engineers, 249, 250, 252
Newcomen Engine, 93, 114-15, 117, 118
Nuclear Regulatory Commission, 246, 248

Pantheon, 21
Parthenon, 11
Patents, 60, 115, 120, 133 n.17, 167, 168, 239
Pont du Gard, 24
Professionalization, 226-27, 235-36, 240-255
Pyramids, 8, 22, 160

Railroads, 99-103, 144, 145-46, 147, 148, 150, 152, 186
Remuneration, 9, 23, 41, 50, 78, 93, 102, 125
Roads, 87, 90, 92, 95
Royal Society, 93, 98, 113

Science and engineering, 2, 14, 52, 64, 71, 79, 88, 119-20, 152, 164, 207, 208, 209, 210, 219-220, 222, 223, 224, 226, 227, 241, 254
Scientific Management, 192-196, 240
Self-employment, 162, 252
Slavery, 19, 26-27, 36
Social backgrounds of engineers, 9, 29, 59, 60, 61, 65, 68, 69-70, 72, 78, 88-89, 91, 92, 97, 99, 101, 105, 114, 116, 120, 122, 124, 125, 126, 127, 129-31, 139, 141, 143-44, 146, 148-49, 153, 171, 226, 236-238
Steam engine, 96, 99-101, 111, 112-122, 124, 126, 141, 210

Technical literature, 60-61, 69, 79, 82 n.27
Technische hochsculen, 212
Technocracy, 196-99
Tennessee Valley Authority, 245-46

Unemployment, 166-67, 198, 217
Unit operations, 241
U.S. Army Corps of Engineers, 86, 143
U.S. Department of Defense, 166
U.S. Military Academy, 142-43, 150, 213, 229 n.24

Water projects, 1, 7-8, 10, 11, 15, 17-18, 19-20, 21-22, 25, 58, 66, 72, 75, 146, 148
Water wheel, 13, 19, 25, 38, 50, 70, 91, 141, 207
Wickenden Report, 188
Women engineers, 238-40

Worcester Polytechnic Institute
Studies in Science, Technology, and Culture

Worcester Polytechnic Institute Studies in Science, Technology, and Culture aims to publish monographs, tightly-edited collections of essays, and research tools in interdisciplinary topics which investigate the relationships of science and technology to social and cultural issues and impacts. The series is edited by Lance Schachterle (Chair, Division of Interdisciplinary Affairs and Professor of English, WPI) and Francis C. Lutz (Associate Dean for Projects and Professor of Civil Engineering, WPI). The editors invite proposals in English from beginning and established scholars throughout the world whose research interests focus on how science or technology affects the structure, values, quality, or management of our society. The series complements WPI's commitment to interdisciplinary education by providing opportunities to publish on the widest possible diversity of themes at the intersection of science, technology, and culture.

TA 15 .R33 1993

Rae, John, 1944-1988.

The engineer in history